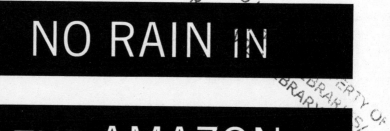

NO RAIN IN

THE AMAZON

ALSO BY NIKOLAS KOZLOFF
AND AVAILABLE FROM PALGRAVE MACMILLAN

Revolution!

Hugo Chávez

NO RAIN IN

THE AMAZON

HOW SOUTH AMERICA'S CLIMATE CHANGE

AFFECTS THE ENTIRE PLANET

Nikolas Kozloff

palgrave
macmillan

First published in 2010 by
PALGRAVE MACMILLAN™
175 Fifth Avenue, New York, N.Y. 10010 and
Houndmills, Basingstoke, Hampshire, England RG21 6XS.
Companies and representatives throughout the world.

PALGRAVE MACMILLAN is the global academic imprint of the Palgrave
Macmillan division of St. Martin's Press, LLC and of Palgrave Macmillan Ltd.
Macmillan® is a registered trademark in the United States, United Kingdom and
other countries. Palgrave is a registered trademark in the European Union and
other countries.

ISBN: 978-0-230-61476-5

Library of Congress Cataloging-in-Publication Data
Kozloff, Nikolas.
    No rain in the Amazon : how South America's climate change affects the
entire planet / Nikolas Kozloff.
        p.   cm.
    Includes index.
    ISBN 978–0–230–61476–5
    1. Climatic changes—South America.   2. Rain forest conservation—South
America.   3. Greenhouse gases—South America.   4. Climatic changes—
International cooperation.   I. Title.
QC903.2.S63K69   2010
333.75'16098—dc22

                                                                        2009033416

A catalogue record of the book is available from the British Library.

Design by Letra Libre

First edition: April 2010
10 9 8 7 6 5 4 3 2 1
Printed in the United States of America.

CONTENTS

ACKNOWLEDGMENTS

I would like to thank Luba Ostashevsky as well as the entire editorial and production team at Palgrave Macmillan, all of the people who took the time to sit down with me in Peru and Brazil, and my grandparents, Leonard and Adele Blumberg of Bound Brook, New Jersey.

NO RAIN IN

THE AMAZON

INTRODUCTION

G lobal warming: For years politicians in Washington, D.C., ig-
nored the issue even as scientists consistently warned about the
imminent dangers of climate change. Now at long last the media
is paying more attention to the phenomenon and the public is becoming
more informed. We've woken up to the harsh reality of climate change,
and not a moment too soon: Already we've witnessed droughts, wildfires,
and increasingly dangerous hurricanes, and we may yet face even greater
challenges to public health. Today's hotter weather allows deadly mosquitoes and tropical diseases to travel farther distances. And humans are
not the only ones who will suffer: Rising temperatures are expected to
disrupt ecosystems and push to extinction those species that cannot adapt.
Rising temperatures will also spur the melting of glaciers and ice caps. An
ominous warning sign of what's to come appeared in 2002 when the
Larsen B ice shelf collapsed in Antarctica.

Global warming is caused by the emission of greenhouse gases, including methane, various nitrous oxides, and carbon dioxide or "carbon
emissions," the most prevalent greenhouse gas produced today. Greenhouse gases are not inherently bad—in fact they occur naturally in the atmosphere and help to moderate the global climate that supports living
organisms, including us. But burning fossil fuels for energy releases large
amounts of greenhouse gases, placing our planet in peril. Since the Industrial Revolution fossil fuels such as oil, gas, and coal have driven the
world's economies. Practically all aspects of our lives—from transport to
central heating—depend on cheap energy, which in turn depends on fossil fuels. When greenhouse gases accumulate in the atmosphere they trap

heat and send it back to the Earth's surface. Over time, this leads to an increase in surface temperature, allowing global warming to spiral out of control.

Before the Industrial Revolution atmospheric carbon dioxide concentrations were approximately 280 parts per million. In other words, if we could have obtained a block of a million molecules of air from the atmosphere in 1750 it would have had 280 molecules of carbon dioxide. Today, however, a block of the same size would have nearly 389 parts per million, and at our current increases of two parts per million each year, we will add 100 or more parts per million of carbon dioxide over the next fifty years. Since the advent of industrialization the Earth's average temperature has gone up between 1 and 1.6 degrees Fahrenheit, and global warming continues at an alarming rate. If we fail to reduce emissions of greenhouse gases the consequences could be dire. Indeed, scientists believe that if we exceed 550 parts per million of carbon dioxide the Earth will face catastrophic climate change.

In the Global North many are not aware that global warming has already exacted a huge toll on people living in the tropics. In this book I look at the cost of global warming in South America, principally through the lens of two large countries—Peru and Brazil. In the former country Andean glaciers are melting at an alarming rate, endangering water supplies. What's more, Peru has been reeling from El Niño, a meteorological phenomenon associated with the Pacific Ocean. In recent years the country's coastline has sustained heavy damage from flooding associated with El Niño, and scientists think that the weather phenomenon may be increasing in frequency and intensity as a result of global warming. A poor country, Peru is hardly up to the task of confronting climate change and has had difficulty coping with tropical disease outbreaks linked to global warming. Meanwhile, across South America Brazil has had to deal with strange, hurricane-like storms linked to warmer ocean temperatures in the Atlantic.

As if Peru and Brazil didn't have their hands full enough they must now contend with an even greater environmental emergency in the Amazon jungle. The Earth's forests remove large quantities of carbon dioxide from the atmosphere and most of this carbon gets locked up in living and decaying vegetation. The world's tropical forests perform a vital function, as they collectively absorb approximately 4.8 billion tons of carbon dioxide annually, and the Amazon is the single largest rainforest sink.[1] Now, however, El Niño has led to intense droughts in the Amazon, which has resulted in forest fires and the release of tons of carbon back into the air.

Indeed, today tropical deforestation has become an important part of the climate change jigsaw puzzle, and currently it's the second largest driver of carbon dioxide emissions. Consider the following sobering statistic: Half of all deforestation in human history has occurred in the second half of the twentieth century, almost all of it in tropical zones inhabited by the world's poorest people. Today, tropical deforestation alone accounts for a full 20 percent of global greenhouse gas emissions.[2] To put it in concrete terms, tropical deforestation releases about 1.5 billion metric tons of carbon into the atmosphere every year.[3] The Amazon contains about one-tenth of the total carbon stored in land ecosystems, and currently the Amazon accounts for nearly half of the carbon dioxide emissions resulting from tropical deforestation. Were the Amazon rainforest to collapse, 26 billion tons of carbon would be released into the atmosphere. To put it in perspective, that's as much carbon as all the nations on the planet produce in the course of two and a half years.[4]

Humanity can ill afford to lose the Amazon River Basin. It's the largest rainforest on Earth and is approximately the size of the forty-eight contiguous United States. It covers 40 percent of the South American continent, including parts of eight South American countries. The rainforest is comprised of a myriad of ecosystems, and vegetation types including rainforests, seasonal forests, deciduous forests, flooded forests, and savannas. The basin is drained by the Amazon River, the world's second longest river (after the Nile). A whopping one-fifth of the Earth's fresh water moves through the Amazon Basin. Perhaps even more startling, the Amazon is home to more species of plants and animals than any other terrestrial ecosystem on the face of the Earth. Today, hundreds of thousands of indigenous peoples live in the rainforest.

Unfortunately, far from helping to arrest climate change in the Amazon, the Global North has exacerbated the problem by promoting destructive industries that hasten deforestation. Already a full 6,900 square miles of rainforest on average gets cleared every year in Brazil. What's causing the destruction? In the tropics subsistence activities have figured prominently in agriculture-driven deforestation, but today's large-scale commercial enterprises are playing an increasingly important role. Within the Peruvian Amazon, logging and coca production have been particularly destructive, while in Brazil industrial-scale cattle ranching and soybean farming are increasingly important drivers of deforestation. When people clear forests for agriculture or other uses they frequently burn a large portion of the aboveground biomass, which releases carbon into the atmosphere. But wait, there's more: Cattle ranching results in methane emissions and agricultural settlers use fertilizer to plant crops, both of which result in nitrous oxide emissions.

If this all sounds a bit removed from your day-to-day life, think again: Brazil exports its tropical commodities to overseas markets in the Global North, and U.S. companies have cashed in on the agribusiness bonanza. What's more, large financial institutions backed by the United States promote climate-unfriendly industries that are encroaching on the rainforest. And if that wasn't bad enough, affluent countries further worsen the environmental situation in the Amazon by promoting energy boondoggles. In Peru, U.S. energy companies have been running roughshod over the rights of indigenous peoples and exacerbating deforestation. In Brazil, U.S. investors promote the use of so-called biofuels, which take up land and push destructive settlers into the rainforest. At the same time large U.S.-backed financial institutions push hydroelectric projects that release potent greenhouse gases into the atmosphere. The same institutions, allied to powerful politicians in Washington, have doled out lucrative loans to the biofuel industry. Needless to say, the affluent countries have failed to provide cleaner, greener technologies to South America so as to head off severe climate change.

To be sure this book criticizes the Global North for ignoring climate change, a phenomenon which stands to have a disproportionately negative impact upon poor nations of the south. Yet, it is much more than a simple polemical attack: I have provided concrete suggestions for coping with climate change in the tropics. In order to understand these serious problems I traveled throughout Peru and Brazil, where I spoke with government officials, experts, environmentalists, and indigenous peoples. From the Peruvian capitol of Lima to the mountain city of Huaraz, to the Amazonian towns of Iquitos and Manaus, local residents repeatedly expressed concern about climate change and its future social and ecological effects. Unfortunately, far too often such concerns go unheard in U.S. media outlets. This book provides a voice to peoples of the Global South while drawing attention to issues of vital environmental importance for us all.

Chapter One

THE GLOBAL NORTH

FAILS TO ACT

You don't need to tell poor *campesino* farmers in Peru about climate change—for years they've been living it. From melting glaciers to glacial flood bursts to impromptu frosts and hail, the weather has bewildered the Peruvian people and they don't know what's coming next. Above all they wonder what's wrong with the Global North. In particular, Peruvians are going to be looking to the developed world to provide more leadership at United Nations climate talks held in Copenhagen, Denmark. Given that the Global North has been polluting since the dawn of the Industrial Revolution, it seems perfectly reasonable to expect richer countries to cut their emissions and provide funding to poorer countries in order to ameliorate the ravages of climate change. In particular, poor countries want affluent nations to commit to cutting their carbon emissions by 25–40 percent of 1990 levels by 2020. Poor countries say they're willing to shift to low-carbon growth, but only if they receive sophisticated technology and funding to make the transition. The United States, which failed to ratify the Kyoto Protocol climate agreement, now says it wants to be part of a new agreement at Copenhagen. However, Washington says it's unrealistic to commit to a 25 percent target in the next decade.

That leaves poorer countries defenseless. As the world grows more and more concerned about climate change and its possible effects, experts are paying particular attention to glaciers, which provide a critical water supply to millions in poor nations. The atmosphere's temperature is rising fastest at high altitudes, and from Alaska to Montana's Glacier National Park to Mount Kilimanjaro in East Africa, glaciers are in retreat. Much of the world's fresh water is in glaciers atop mountains. The glaciers act as gigantic storehouses: In wet or cold seasons, they grow with snow; in dry and hot seasons, the edges slowly melt, gently feeding streams and rivers. More practically, they are a vital part of the planet's system for collecting, storing, and delivering the fresh water that billions of people depend on for washing, drinking, and agriculture. Farms are dependent on glacier meltwater; huge cities have developed on the belief that mountains will always give them drinking water; and hydroelectric dams rely on the water flow to generate power.

In Asia and Latin America, hundreds of millions of people use runoff from glaciers for irrigating crops, washing, and drinking water. In the Andes, so-called "tropical glaciers" spread out over 965 square miles and form an imposing landscape. Twenty percent of the region's tropical glaciers are in Bolivia, while 8 percent lie in Ecuador and Colombia.[1] Scientists are particularly concerned about the rate of ice loss atop these tropical glaciers. According to experts, ice loss is actually accelerating which represents a significant problem as with rapid loss the ice cannot replenish itself.

Although a few glaciers in southern Patagonia are increasing in size, almost all located near the tropics are undergoing rapid retreat. Particularly hard hit is Colombia, where some glaciers are now less than 20 percent of the mass recorded in 1850. In 1983, the five most important glaciers in El Cocuy National Park were expected to last for another 300 years at least. But according to measurements taken in 2005, these glaciers could vanish within twenty-five years. Meanwhile, the ice sheet on the Ecuadorean volcano of Cotopaxi and its glacier has retreated by 30 percent since 1976. Overall, Ecuador could lose half of its most important glaciers within twenty years. In Venezuela's provincial state of Mérida, only a small number of glaciers remain.[2] In Bolivia, the Chacaltaya glacier is expected to completely melt within fifteen years under present conditions. In Chile, the O'Higgins glacier has shrunk by nine miles in one hundred years, and in neighboring Argentina the Upsala glacier is losing fourteen meters a year.

Peru contains a full 71 percent of South America's tropical glaciers, and to the dismay of many local inhabitants ice peaks are now turning brown. Quelccaya, the world's largest tropical ice cap, is retreating at about 200 feet a year, up from 20 feet in the 1960s. Over the past few decades Peru has lost 22 percent of its glacier surface area. That spells trouble for the Andean nation, which relies on glaciers for much of its water supply. According to the authorities, the country has lost seven billion cubic meters of water as a result of glacier melt. That's the same amount of water consumed by Lima, a city of more than eight million people, over the course of ten years. In the early 1980s, when you traveled between the towns of Huaraz and Chavín you could see impressive glaciers. Passing by the Chanicocha Lagoon, onlookers would marvel at an enormous peak covered in snow year round. Now, however, the snowy area has been reduced by 90 percent.

In an era of global warming, certain Peruvian glaciers have fared particularly poorly. Take, for example, the case of the Broggi glacier. Old black-and-white photos of Broggi from the 1930s show the glacier full of ice. By 2005, however, the glacier had entirely disappeared. Within the Andes' so-called White Mountain Range, 145 small glaciers disappeared between 1970 and 2003. The Black Mountain Range, located near the White range, surpasses 5,000 meters in altitude. These mountains have lost all their glaciers. One of Peru's most celebrated glaciers, Pastoruri, has also been melting at an accelerated rate. In the mid-1970s one could hike up the glacier. By the early 1980s, however, observers noted that Pastoruri had less snow and the peak's well-known ice caverns were disappearing. Ten years later hikers could not find their treasured caves anymore.[3] In early 2007 the last cavern on Pastoruri collapsed under an unusually strong Andean sun. Tourist postcards showing the great caverns of Pastoruri have become obsolete.[4]

Scientists are also concerned about another glacier, Quelccaya. Before it melts, scientists are determined to unlock its climate secrets. Researchers have long argued that ice is a precious time capsule that can reveal crucial climate swings that occurred over the course of history. For example, fluctuations in various isotopes, or atomic forms, of oxygen chronicle shifts between warm and cold periods. At the same time, fluctuations in nitrate levels indicate how plants responded to the expansion and contraction of ice. Ice contains air bubbles from the ancient atmosphere and ash layers from volcanic eruptions long ago. It also has layers of windblown dust that reveal key information about broad shifts in precipitation; dust rises during dry periods and falls during wet epochs. What's more, ice records shifts in rainfall in the form of thicker or thinner annual layers.

After they extracted ice cores from Quelccaya, scientists were able to ascertain an intriguing 1,500-year-long climate record. By analyzing the cores, researchers were able to piece together the droughts and floods that unhinged pre-Incan civilizations. Scientists saw dramatic swings from wet to dry that coincided with variations in sea-surface temperatures typical of the El Niño weather phenomenon. In addition they discerned more long-term shifts, from rainy periods to droughts that lasted decades and even centuries. Archaeologists studied weather patterns that eerily paralleled the rise and fall of a great pre-Incan civilization called Tiwanaku that flourished in the area of Lake Titicaca more than a thousand years ago.

Sifting through the ice cores, researchers saw that they appeared to reveal a wavelike sweep of ice growth proceeding south to north across the Equator. The pattern bore a marked correspondence to a 21,500-year astronomical cycle known as the precessional cycle. Like a child's top, the Earth wobbles as it spins which changes the time of year in which the Northern and Southern Hemispheres make their closest approach to the sun. That in turn affects rainfall patterns. According to experts, the precessional pattern is still at work but its influence has become harder to detect. "To me this is what makes our world today seem so different from the past," says glacier expert Lonnie Thompson. "If nature alone were in charge, then glaciers should be growing in the lower latitudes of one hemisphere and retreating in the lower latitudes of another. But that's not what's happening."

As Thompson sees it, the fact that glaciers are vanishing all across the planet constitutes a clear sign that the greenhouse effect is damaging the natural system. By drilling ice cores through to bedrock, extracting samples, and periodically monitoring the slow but accelerating retreat of Peruvian glaciers, he has amassed vital evidence of climate change: As the ice retreats, ancient plant beds have been newly uncovered. Carbon dating indicates that most of these plant beds have been buried for at least 5,000 years. The current retreat of the ice exceeds any other retreat in at least the last fifty centuries, researchers claim. By analyzing these ice cores, scientists have concluded that temperatures throughout the tropics are increasing.

Thirty years ago, the peak of Quelccaya was a dazzling expanse of white. Some 18,700 feet high, the huge glacier in Peru extended over 22 miles. The dramatic western face of Quelccaya looked remarkably like a 180-foot-high wedding cake. Along the face of the glacier, scientists noted layers of ice alternating with dust. If the cliff face ever melted, researchers believed, the sharply delineated layers would collapse into homogenized slush. In just thirty-three years, however, Quelccaya has shrunk 30 percent. When discussing disappearing peaks, Thompson frequently draws an analogy between the proverbial "canary in a coal mine" and glaciers. Like the canary, he says, glaciers are warning humankind of the buildup of dangerous gases. There is one important difference, however: "In the past," he says, "when the canaries stopped singing and died, the miners knew to get out of the mine. Our problem is we live in the mine."[5]

If glaciers disappear this could hit Peru's tourist and mountain climbing industry particularly hard. In Huaraz the proprietor of my humble hotel expressed grave concern about glacier melt. However, he had not met with tourism officials in the government nor was there a long-term economic plan afoot to help Huaraz's tourist industry confront the potential challenges ahead. Huascarán, located east of the town of Yungay, is Peru's most famous mountain. It is the highest peak in the country and a favorite among mountain climbers and tourists alike. Situated within the White Mountain Range, it has deep ravines watered by numerous torrents, glacial lakes, and a variety of vegetation, which makes it "a site of spectacular beauty."[6] But Huascarán is faring even worse than Quelccaya: The peak has lost about 40 percent of its ice area over the past thirty years.

The White Mountains possess the last remaining dwarf high-altitude clouded forests and Huascarán is home to such species as the spectacled bear and the Andean condor. For millennia, indigenous cultures have flourished around Huascarán, and dozens of archaeological sites are known throughout the area. According to legend, Huascarán and another mountain in the White Mountains Range, Huandoy, came into existence as a result of divine intervention. The Indians believe that there were once two empires that were bitter enemies and for years there was constant war. The kings of both empires had children: One king had a son named Huascar and the other king had a daughter named Huandy. Just like some kind of ancient Andean Romeo and Juliet, the two youngsters fell in love. But the couple was trapped: If they continued to love one another, they knew their parents would discover their illicit relationship and that it could lead to yet another bloody war. Keen to avoid conflict, the two lovers were careful to meet only at night when they would be safe from disapproving eyes. Tragically, word reached both kings of the affair and the couple was discovered. Forced to flee, the lovers were later caught and punished harshly. Chained face to face to a mountainside, unable to embrace each other, the couple cried endlessly—so much that their tears turned into rivers and froze to ice. Moved by the lovers' suffering, God showed mercy upon the couple and changed them into mountains: Huascar was turned into Huascarán and Huandy into Huandoy. To this day, it is said, Huascar and Huandy continue to cry and their tears flow into each other's lakes.[7]

The disappearance of Huascarán and other snowy peaks would prove difficult for many Peruvian Indians: For them, the mountain is a symbol of cultural pride. Ever since the days of the Inca Empire thousands of people have made an annual pilgrimage known as Qoyllur Rit'i to the mountain of Ausangate in the month of June. A peak located in southeastern Peru, Ausangate is sacred to the Quechua-speaking Indians. During the pilgrimage and festival, dancers stand all night on top of the 15,000-foot glacier.[8] Traditionally, members of the procession stand at the glacier so as to have the privilege of taking a chunk of ice from the mountain. The pilgrims take back blocks of ice to bless their lands and crops. They also give thanks to the *apus*, sacred mountains and the most powerful earthly deities, for their generosity. Later the community melts the ice and drinks the water, which is considered sacred.

In 2007, however, it was impossible to carry out the ritual since there was so little snow on Ausangate. For local Quechua-speaking Indians the changing physical landscape is culturally disorienting. Indigenous people refer to prominent features in the landscape as *tirakuna* or "the ones who watch over us." Within indigenous cosmology nature plays a central role, and though Indians are Catholicized, the Virgin Mary is identified with Pacha Mama or Mother Earth. When indigenous peoples drink alcohol they always make sure to pour a drop on the ground for her first.[9] In 2007, out of respect to the *apus* the *ukukos* (members of the Qoyllur Rit'i procession) decided not to take any ice from Ausangate. Alarmed, some farmers interpret the disappearance of their glaciers as foreshadowing the end of the world.[10]

In addition to Huascarán, climate change stands to unleash environmental chaos upon one of Peru's most beloved archaeological sites: Machu Picchu. Standing at 2,430 meters above sea level in the middle of a tropical mountain forest, Machu Picchu lies in an extraordinarily beautiful setting. Pachacutec Inca Yupanqui, a fifteenth century Inca ruler, constructed Machu Picchu with the help of his descendants. At its zenith, the site was probably the most astounding urban creation of the Inca Empire. The city's giant walls, terraces, and ramps appear as if they were cut naturally within continuous rock escarpments. Machu Picchu has been declared a World Heritage Site and one of the Modern Wonders of the World. Since its discovery in the early twentieth century, Machu Picchu has been inundated with tourists. In 1992 Machu Picchu received 9,000 tourists, but by 2006 the figure had jumped to 4,000 *on a single day.[11]*

Though many tourists might not be aware, Machu Picchu is at ecological risk. In 1998, the nearby Salkantay glacier collapsed and a huge landslide buried a hydroelectric power station. Water gushed through the valley below Machu Picchu and flooded the area. Fleeing inhabitants, with nowhere else to go, climbed up on the slope of a mountain while others perished. Rescuers from Machu Picchu, using antiquated Russian helicopters, helped to gather up the traumatized local Quechua Indians and *campesino* peasants. The survivors had lost everything including their homes and possessions. Below the rescue site the whole area had become flooded, creating a gigantic lake. Relief workers helped to evacuate the survivors to Aguas Calientes, a nearby town. The hydroelectric power station was shuttered for a year and reconstruction cost the government $400 million.

While small landslides were routine in the area, what happened in 1998 was different. What witnesses saw in and around Machu Picchu is known in Peru as an *aluvión*, a glacial flood burst, and people have grown to fear such events. Experts point out that 1998 marked the appearance of the weather phenomenon known as El Niño, and that rising temperatures helped to melt the ice on the peaks surrounding the hydroelectric station.[12] In addition to melting glaciers, local residents in Cusco and the Urubamba Valley are concerned about changing rain patterns. In the short term, Peru could experience cycles of extreme rain that fall within a short period of time. This in turn could create more landslides and flooding, similar to what occurred in Machu Picchu. Paradoxically, however, the country might also experience long periods of prolonged drought.

Machu Picchu will be right in the crosshairs of delicate climate change for the foreseeable future. Every summer in Peru (the opposite of U.S. winter) the country experiences a rainy season in the south. The Andean nation is hit by a series of *aluviones*, resulting in mudslides that destroy houses and block highways. *Aluviones* occur periodically when water is released abruptly from a previously ice-dammed lake alongside, within, or above a glacier. Such terrifying flood bursts typically arrive with little or no warning, carrying liquid mud, large rock boulders, and blocks of ice. Towns lying near glacial outwash streams are particularly vulnerable to such catastrophic flooding. When it's a peak El Niño year, like 1998, the number of *aluviones* goes up and the resulting disruption for people and infrastructure is correspondingly greater. Concerned about the threat, authorities have improved urban planning and zoning so as to prepare for *aluviones*. And yet, Machu Picchu remains woefully unprepared for another disaster.

When traveling to Machu Picchu by train one stops off in Aguas Calientes, also known as Machu Picchu town, which is known for its springs and natural thermal waters. From there tourists proceed to the archaeological site of Machu Picchu by bus. The town itself is constructed near a creek that feeds into a tributary of the Vilcanota River. In a peak year Aguas Calientes could be inundated by an *aluvión*, and yet local residents stubbornly refuse to move. Despite warnings from local and national officials as well as the World Bank, townspeople are reluctant to give up their livelihoods. Hotel and restaurant owners are aware of the risk of an *aluvión*. But, in the short term they figure they're making a lot of money as a result of the tourist boom at Machu Picchu.[13]

Like Machu Picchu, the town of Huaraz has had a long and terrible history of glacial disasters. In 1941 an *aluvión* destroyed one-third of the town, killing 5,000–7,000 people in the process. The catastrophe prompted the government to get into the business of glacier research by hiring a group of scientists to study the problem. The authorities constructed drainage systems in lagoons to help in the event of another *aluvión*. In 2003 glacier ice fell into Lake Palcacocha, cutting off the water supply to Huaraz's 60,000 residents for eight hours. In October 2008, after spending some two weeks in Lima, I headed to Huaraz to see the glaciers for myself. The town is located six hours by bus from the capital along a long and desolate mountain highway.

Marco Zapata is the coordinator of the government's glaciology and hydraulic resources unit in the town. A veteran scientist, he arrived in Huaraz in late 1970 shortly after a devastating earthquake hit the town. The department of Ancash (of which Huaraz is the capital city), Zapata explains, has had more natural disasters, including avalanches and *aluviones*, than any other area of the country. Glaciers tower over Huaraz and tourists flock to the town to go mountain climbing. "People used to think that the glaciers were eternal, and didn't pay much attention to them," says Zapata. But scientists, who have been measuring climate change in Peru for some thirty years, believe glacier retreat in the area is accelerating and could be irreversible.[14]

Other signs of ecological confusion abound. For example, local residents are now discovering mosquitoes, flies, and even butterflies for the first time at the base of glaciers.[15] Zapata is concerned that tropical disease might spread to higher altitudes. Meanwhile, plants will also have to adapt to climate change. During our interview, I asked Zapata about the cactus in front of his office. "Cacti come from a more tropical zone but they're adapting quite well to Huaraz," he says.[16]

Scientists are acutely aware of the human toll that glacier melt will exact on everyday Peruvians. For example, farmers in the area of the Santa River who grow potatoes, wheat, and artichokes depend on glacier melt from the White Mountain Range during the dry season. If the glaciers dry up then literally so will the farmers' livelihoods. In the Santa River valley, where glacier melt has exposed bare mountain peaks, water has become scarce and farmers aren't hiring any more hands. What's more, pack animals, which the farmers depend on, are also being affected by climate change and water scarcity. Alpacas normally drink from nearby streams. But now they are falling ill after drinking stagnant and muddy water. The Andes are home to many rare species in addition to alpacas, such as condor, pumas, and llamas, all of which will come under threat when the water supply disappears.

As if farmers did not already have enough climactic worries, the disappearance of local glaciers could have even more dire consequences. Glaciers function as shock absorbers for rain, accumulating it in the wet season and discharging it in the dry season. But they also help pry precipitation loose from the moist air that rises up from the Amazon basin below.[17] So if glaciers disappear, this could result in less rain. Andean *campesinos* can only hope that whatever water they lose from glacier melt will be compensated for in other ways. Zapata believes that global warming will result in more evaporation and produce more rain, potentially bringing some relief.[18]

Glacier melt stands to have long-term repercussions on Peru's economy because the icy peaks generate 70 percent of the country's electrical power, and current boom sectors such as mining absorb huge amounts of water. Glaciers are tremendously important in meeting the country's water needs: Two-thirds of the country's 27 million people live on the bone-dry coast where just 1.8 percent of the nation's water supply is found. Andean glaciers feed the rivers that lead to sprawling cities and shantytowns on Peru's arid Pacific coast. In the short term, melting glaciers could represent a water bonanza for Peru. But in the long term, disappearing glaciers could signify a loss of seven billion cubic meters of fresh water for Lima and coastal towns. The thirsty capital is a particular worry. Lima is built on a desert. The city supports a population of more than eight million and receives hardly any rainfall. With thousands of new arrivals every year, the city's water demand is only set to increase. It's a worrying prospect for urban planners who still remember a disastrous drought in 2004 that pushed Lima's water supplies to their limit.

Just twenty years ago the strip of land between the Pacific Ocean and the Andean foothills was bare save for an occasional fig or carob tree. Today, however, this barren desert is home to Peru's dynamic agro-export sector, which cultivates vegetables such as asparagus. Where is the water going to come from to service all the burgeoning farms? That is the pressing question on many people's minds as rising demand for irrigation drains Peru's water supply. Approximately 80 percent of Peru's water goes toward agriculture, but water conserving systems like drip irrigation are underutilized. As Andean ice caps disappear the country will have to move quickly toward more efficient irrigation. For Peruvians the water crisis is looming and it could pit different regions of the country against each other. One scheme afoot would channel water from the Andes to the coast. The plan, however, could lead to conflict between big agribusiness and Quechua-speaking llama herders in the mountains who rely on spongy high mountain wetlands for pasture.

While experts say that water shortages could be offset by expanding reservoirs set up to catch water during the rainy season, it is unclear what kind of impact the loss of glacial runoff could have upon water supplies downstream. Indeed, as cities and agriculture expand in Peru and throughout South America, climate change will exacerbate water scarcity, increase conflicts between rival users, and pit city dwellers against rural farmers. By 2020 tens of millions of South Americans could be facing water shortages, an unfortunate trend that is likely to carry over to the rest of the world. In Lima, some governmental authorities are taking the glacier issue seriously. In the new Environment Ministry, Laura Avellaneda tells me that she and her colleagues are studying the effect of melting glaciers on water supplies. She adds, however, that they need more researchers.[19] Officials meanwhile say they will try to discourage cultivation of crops that consume large amounts of water.[20]

Despite these positive positions, there's some doubt whether Peru's current president, Alan García, is the man to tackle global warming. "We should not pay attention to those who consider melting and disappearing glaciers a threat," he has said.[21] As long as Peru's glaciers continue to melt, García says, the country may turn the climate crisis to its advantage by selling its excess reserve water to Brazil. Speaking to business executives in São Paulo, García added, "The water will never stop trickling down." The president thinks climate change could lead to an economic bonanza as Peru provides Brazil with abundant hydroelectric energy. But while it's possible Peru may possess excess water from melting glaciers, in the long term the outlook is bleak: Experts say that the Andean nation will run out of water reserves by 2050.[22] Never fear, says García: Peru can always turn to desalinization plants on the coast to solve the country's looming water shortage problem.[23]

With such uninspiring environmental leadership, Peruvians hope that richer nations will help provide welcome financial resources to cope with the ravages of climate change. Indeed, the Global South has beseeched more affluent countries for so-called "adaptation funds," which would go toward mitigating the impact of global warming. Given wealthier countries' responsibility to clean up the mess, it's certainly a reasonable request, and during recent climate negotiations in Poznan, Poland, the United Nations agreed to create such a fund. Those nations that have ratified the Kyoto Protocol climate change agreement contribute to the adaptation fund. Needless to say, the United States has failed to sign the Kyoto agreement, which the United States claimed did not go far enough in making demands of developing countries such as China.

Perhaps the Americans will become more proactive at long last at a climate summit in Copenhagen, Denmark, that is supposed to yield a new climate change treaty. Meanwhile, as the wealthy nations delay, poorer countries are still waiting for their adaptation funds. Unfortunately, none of the money has materialized, and, to add insult to injury, the Europeans have refused to say how much they will commit toward bailing out poor countries. In order to deal with the ravages of climate change, World Wildlife Fund says the Global North needs to immediately start paying $2 billion a year to the Earth's poorest nations. If governments drag their feet, concerned citizens in affluent countries are going to need to exert more pressure so that countries like Peru get a fair shake.

While wealthy nations dither, Peru is trying to sort out its bewildering climate problems. Indeed, when it comes to climate change, melting glaciers may be just the tip of the (disappearing) iceberg for farmers in the Andes. Residents of the community of Vicos in Ancash are feeling the pinch these days. Local residents cultivate traditional crops such as potatoes, corn, beans, peas, and an ancient Inca grain called quinoa. They're particularly concerned about a new fungus that appeared ten years ago that might be linked to abrupt climate change in the mountains.[24] Farmers say that warmer temperatures are paving the way for the emergence of late blight, a disease that caused the Irish potato famine in the mid-1800s.[25]

Late blight is caused by the fungus-like *Phytophthora infestans*, and spores carried in the wind or in infected tubers brought to new regions can result in infestation. At first a few grayish specks appear on the plant's leaves, followed by a cottony film. The disease can easily lead to the destruction of whole fields of potatoes in the event of high humidity or cool to warm temperatures. Late blight can also make a crop unsuitable for future storage. Because of the disease, farmers have had to farm potatoes at higher and colder altitudes. Though the *campesinos* get a better yield this way, it's highly inconvenient for them as they have to travel farther up the mountain.[26] Climate change is bringing new and more frequent diseases during harvest time, say experts. As plagues and temperatures increase, farmers will be obliged to move higher and higher up the mountainside to avoid the new conditions. Eventually, however, they'll have no place else to go. Big losers could include 1.8 million potato farmers who depend on predictable climate. These *campesinos* are ill prepared to deal with new pests and diseases that have emerged as temperature and rainfall patterns start to change.[27]

Late blight could be particularly damaging for Peru: In local villages the potato represents more than just food—It's a culture. Indeed, there are songs, dances, and ceremonies about the potato, and even a potato god. For everyday *campesinos* the potato, which originated in Peru and fed the Inca Empire, is as important a staple as rice to the Chinese. According to the International Potato Center, a single medium-sized potato has about half the daily adult vitamin C requirement, whereas rice and wheat have none. In addition the potato is rich in vitamins B1, B3, and B6 and minerals like potassium, magnesium, and phosphorus. It is also extremely low in fat. The potato has only one-fourth the calories of bread, and boiled it has more protein than maize with nearly twice the calcium.

Experts point to genetic studies demonstrating that all potatoes originated more than 10,000 years ago from one ancestor, *Solanum brevicaule,* which is common to the Peruvian side of Lake Titicaca. Today, there are 3,000 varieties of potato worldwide; 2,000 are found exclusively in Peru.[28] Over the centuries, Peruvian farmers have pointedly sought out and nurtured thousands of different potato varieties because the El Niño phenomenon always brought weather changes. The *campesinos* must adapt their crops to volatile conditions and the diversity of crops they created is a response to the chaos of the system. Today, Peru is hoping that this great genetic diversity will help it survive the ravages of even greater climate change. Protecting potato farmers is a vital economic concern for the country: According to the United Nations Food and Agriculture Or-

ganization (FAO), potatoes are cultivated on 600,000 small farms covering a total of 260,000 hectares.[29]

High in the Andes, farmers are growing dozens of different kinds of
potatoes in the hope of finding the ones that are most suitable to withstand climate change. Though some types of potatoes have been lost over
time, farmers may obtain different varieties of the tuber from a gene bank
at the International Potato Center in Lima. The geneticists offer farmers any type of potato they need. The facilities are full of dry seeds, racks
of test tubes holding plant shoots, and freezers storing germ plasm.[30]
Walking through the center, visitors are greeted by centuries of food history displayed on trays. Potatoes of all colors and shapes—blue, purple,
yellow, red, gold, russet, speckled, round, clustered, long, oval, pear-
shaped—are laid out on tables and are identified by their traditional
names in the indigenous Quechua and Aymara languages.[31]

Today in the village of Paru Paru, farmers are testing some of the institute's potatoes that have been boiled, fried, mashed, and poached for
hundreds of years.[32] In late 2004, six communities in the Sacred Valley of
the Incas created a 10,000-hectare park that is covered with *layme* or
muyuy—communally worked fields—interspersed with ponds and streams.
In total, Potato Park is home to 1,200 families. Now, *campesino* men and
women from Paru Paru are constructing a community greenhouse for the
preservation of seeds.[33] At village gatherings, two young boys summon
everyone by blowing into huge seashells brought up from the Pacific Ocean.
Farmers sit down for a hearty lunch: potato soup mixed with quinoa and
meat. One of the cooks remarks that she has 250 types of potato to choose
from. No doubt, there will be more gatherings like this in Paru Paru to discuss the new weather and which crops grow well in the vicinity. Most likely,
each meeting will commence with a bowl of potato soup.[34]

Meanwhile, within the Agriculture Ministry experts are trying to
determine which traditional tubers are most suitable to withstand the
ravages of global warming. The future may lie with traditional crops
that farmers are already familiar with such as *ollucos*. A yellowish tuber,
olluco was domesticated by pre-Incan populations. However, it remains
to be seen whether exporting *ollucos* or other traditional tubers could

be a viable enterprise, and farmers desperately need more capital and technical know-how.

At this point government must pay more attention to potato farmers and the highlands if it wants to head off looming social problems. Already, people are leaving the Andes because local climactic conditions aren't allowing farmers to survive and stay on their land. In an effort to boost the international profile, production, and trade of the potato, the United Nations Food and Agriculture Organization declared 2008 the International Year of the Potato. Meanwhile, the Peruvian authorities see the potato as a means of alleviating poverty and seek to increase internal and external production. The government, moved by research demonstrating that potato yields more calories per pound than increasingly expensive grains, is carrying out a campaign to replace white bread made from wheat with potato bread in schools.

The rest of the world can ill afford to see important potato varieties wiped out in Peru. In an era of climate change and food crisis, we need to preserve the genetic diversity of our crops as best we can. At long last, wealthier nations have stepped up to the plate and have offered special funding to Quechua Indians and other poor farmers to safeguard threatened crops. Under the plan, Peruvian farmers will get paid to look after their diverse collection of potatoes. It's definitely a good insurance policy against climate catastrophe—in the event that today's commercially available potatoes should fail, we'll be able to rely on Peruvian potatoes that are being farmed at different altitudes and under different climatic conditions. Under the International Treaty on Plant Genetic Resources for Food and Agriculture, affluent nations have contributed more than $100 million to the project. Already, the fund has led to the creation of an international vault containing 1.1 million seed varieties. Unfortunately, the United States has yet to sign up for the fund.

For Peruvian farmers taking advantage of the country's genetic diversity may be a wise move. Indeed, if poor *campesinos* in the Andes have learned anything in recent years, it is that the weather has become wily and unpredictable. Farmers in the village of Vicos have been startled by impromptu frosts and hail. Before, they say, the frosts would occur every three or four years in December or November. But now the frosts can pop up at any time, ruining crops.[35] Meanwhile, extreme cold accompanied by frost and hail (known in Peru as *friaje*), has resulted in temperature drops below 35 degrees Celsius. Such temperature shifts affect poor communities which live off of subsistence agriculture. Incredibly remote and isolated, these communities are located between 4,000 and 4,500 meters above sea level.[36]

When the blizzards came they were strong, local residents say. The snow fell for an entire day and a full night without ceasing. When it finally stopped snowing, the skies opened and it became completely clear. But then ice fell from the sky in huge glass-like shards and the cold front hit people hard.[37] Though cold fronts are nothing new in the Andes, their occurrence has now become more frequent, abrupt, and extreme. In 2004, recurring icy cold fronts accompanied by hail destroyed crops, ruined pastures, and led to the untimely death of cattle in poor and mountainous southern Peru. After temperatures plummeted 50 children died and 13,000 people were left suffering from severe bronchitis, pneumonia, and hypothermia. Needless to say, the snow killed off all vegetation and alpaca died in the thousands.[38] Today, farmers say that the days are hotter but nights have become colder. In February 2007 unseasonably harsh frosts and hail again destroyed crops in the department of Huancavelica in the central south region of Peru, affecting more than 40,000 *campesino* families.[39]

For farmers, losing pack animals like alpacas can prove economically devastating. The Andean camel is a relative of the llama and provides milk and cheese that contain important nutrients. Alpaca wool is dense and provides exceptional insulation. Excess wool is sold to pay for other expenses, such as children's schooling. Without alpacas, *campesinos* would have no way of getting goods like potatoes over mountainous terrain. Moreover, farmers need alpacas to bring medicine and food back to their villages. At the same time, manure constitutes necessary fuel for heating and cooking.

When exposed to excessive cold, alpacas struggle to find food in the snow and ice. Pregnant females can miscarry and survivors wind up exhausted and susceptible to disease. In Pucurami, a small community in the Andean foothills, alpacas are in danger from climate change. The town lies in the vicinity of the life-giving Ausangate glacier, which has been receding in recent years. Loss of snow means that farmers receive less water, and that in turn means less pasture and more problems raising livestock. Alpacas aren't eating enough so their wool doesn't grow as well. As a result, people are forced to turn to synthetic wool to weave hats, sweaters, and scarves.

Today, *campesinos* are taking proactive measures to protect their flocks: They've joined with a British charity called Practical Action to build shelters for the alpacas.[40] The shelters are made from local materials and protect the alpacas, in particular the young and weak, from extremes of cold weather. Housing up to fifty alpacas, the shelters not only protect the animals from wind and cold in winter but also provide space for shearing alpacas and sorting wool in summer.[41]

Meanwhile local farmers are bracing for the upcoming *friaje*. Near farms, black scars mar local paths. *Campesinos* seek to ward off lightning from electric storms and believe that by burning tires on the paths they can protect themselves. Others shoot fireworks into the clouds, hoping they will push bad weather away. To preserve warmth, farmhouse rooms are claustrophobically small. Inside, visitors are greeted by the overwhelming scent of alpaca skins. The farmers hang the dried skeletons of alpaca embryos from their thatched roofs as an insurance policy against poor harvest. It's a tradition that has remained unchanged for hundreds of years. Outside, however, farmers keep solar-powered satellite dishes to avoid being cut off from the outside world in the event of a cold snap.

Getting adequate nutrition at these high altitudes can be a challenge. One food source is the vicuña, a rare deer-like animal. But the Quechua Indians only hunt them once a year so as to maintain a sustainable population of the animals. In the forbidding mountains, cold snaps dry up the land and vegetation simply blows away. Facing a challenging climate and a dwindling food supply, local communities are turning to hydroponics systems, a technology that dates to pre-Christian Rome. In hydroponic cultivation, plants are nourished by nutrient-rich water rather than soil in a carefully controlled environment. In the Andes, farmers are fetching barley grains with the help of alpacas and growing them in troughs of water. The *campesinos* then mill the barley, enrich it with syrup, and form it into blocks. The whole process takes only two weeks and requires only simple sunlight and water. When no other food is available to the alpacas, the high energy blocks of barley keep the animals strong and healthy. Meanwhile, *campesinos* are training to become *kamayoqs:* farmer-to-farmer trainers who pass on knowledge of hydroponics and basic veterinary skills to others. In the Andes this has become vital as many farmers don't know how to protect alpacas from disease. Previously, when animals got sick farmers would have to take them into town, which would take at least a day. While the *campesinos* were away, more animals might become sick, as disease spreads rapidly.

As human and animal populations struggle to adapt to global warming in the Andes, changing weather patterns spell concern for Andean cloud forests. These forests, together with the adjacent Amazonian lowlands, form the most biodiverse region on Earth. Uniquely, such forests capture moisture through condensation from the clouds. Cloud forests are particularly vulnerable to climate change—indeed if temperatures increase merely one degree in the lowlands this translates into big trouble in the mountains, with a rise of two degrees.[42] According to experts, rising temperatures are causing clouds that blanket the Andes to condense at higher altitudes. Eventually the so-called dew point will miss the mountains altogether and the cloud forest will dry out.[43] If you think this is merely a local Andean concern you are dead wrong. As the cloud forest seeks to adapt to changing temperatures it may move upslope into grasslands, which are subject to frequent fire. Typically, cloud forest stores more carbon in dead biomass than in branches and leaves. When wildfire strikes the carbon will get released to the atmosphere very rapidly.

If Andean cloud forests disappear the Earth will have lost a great marvel. To get a sense of the truly dramatic topography and climate variation in Peru, head from the snow-capped mountains to the treeless plains and dry valleys of the *altiplano* before suddenly descending.

Eventually you arrive in the cloud forests, which act as the headwater region for the Amazon. These fragile worlds display incredible variation in altitude, slope, aspect, exposure, and geological substrate. A great variety of cloud forest trees are supported here by moisture rising up from the Amazon. The trees in turn sustain ferns, mosses, bromeliads, and orchids, which take advantage of any spare patch of bark to lay down their roots. It is precisely these so-called epiphytes ("epi" means "on top of" and "phyte" means "plant")—in addition to wet humus soil, the thick understory of plants, and the heavy cloud cover—that make cloud forests unique. The atmosphere is damp and cool, and during much of the day an eerie mist hangs over the forest and bird calls carry vast distances. When the sun succeeds in breaking through it filters through the leaf canopy and highlights the intense color of local flowers and foliage.

Cloud forests are located in tropical and subtropical mountainous regions where cooler temperatures prevailing on mountain slopes cause clouds to form. They comprise a little less than 400,000 square kilometers, or less than 2.5 percent, of the Earth's tropical rainforests. Currently, 60 percent of cloud forests are located in Asia while about 25 percent lie in Latin America. In Central and South America cloud forests stretch all the way from Panama to Argentina. Typically encountered between 5,000 and 10,000 feet above sea level, cloud forests carry out a vital hydrological function as trees strip water from windblown fog and clouds, thus helping to sustain neighboring ecosystems.

Cloud forests are also repositories of extraordinary life: The Centinella Ridge in Western Ecuador, for example, contains approximately ninety endemic plant species within a forest area of merely twenty square kilometers.[44] Head to the Venezuelan mountain summit known as Cerro de la Neblina and you can find shrubs, orchids, and insect-eating plants, almost all of which are found no place else. In this sense cloud forests act as habitat pockets where different species are found on each mountain range.[45] Cloud forests also serve as important gene pools for crop improvements as they harbor wild relatives of staple crops such as papaya, tomato, passion fruit, avocado, beans, blackberry, cucumber, potato, and peppers.

Furthermore, new species are frequently discovered in cloud forests. In the 1990s, for example, researchers found a novel bird in the Ecuadoran cloud forest called the Jocotoco Antpitta.[46] Peru's cloud forests are also full of bountiful wildlife—indeed over 30 percent of the country's 272 species of endemic mammals, birds, and frogs are found in these ecosystems.[47] Perhaps many species are still yet to be found. In 1999, researchers discovered a strikingly unusual large rodent. Approximately the size of a squirrel, the creature is nocturnal, has long dense fur, a broad block-like head, and a thick furry tail. On its crown, nape, and shoulders the rodent has a blackish crest of fur that adds to its unique appearance. Researchers discovered the rodent at 6,200 feet but later attempts to find and observe the creature proved unsuccessful.[48]

How will cloud forest wildlife "weather" climate change? Recent developments have not been promising. In 1987 El Niño hit Costa Rica, resulting in several weeks of dry weather in the Monteverde cloud forest. Some researchers have suggested that El Niño may have been particularly strong that year because of global warming. In the wake of the dry spell, a full twenty-five of the fifty frog and toad species in the area simply disappeared. Only five returned later.[49] Hopefully history will not repeat itself in the Andean cloud forest, which harbors many species of frog including the recently discovered Noble's Pygmy frog, so small it could sit on a dime with room to spare.[50]

In addition to wildlife, cloud forest plants are also in danger. Because cool to cold temperatures and poor soils prevail in the cloud forest, plants grow slowly. With temperatures rising up to five degrees Celsius by 2100, most of these plants are not expected to survive.[51]

Frogs are not the only creatures that depend on the cloud forest. If this vital ecosystem were to collapse it would affect people all across the Andean region. That's because cloud forests act as giant sponges that absorb rain, condense clouds, and gradually release water into the watersheds below. As nature's "water towers," cloud forests provide billions of gallons of fresh water to downstream industry, towns, and regions. Moreover cloud forests act as the headwater regions where most Amazonian migratory fish spawn. If the cloud forest goes belly up, the region will have big problems, since fish provide most of the protein for human consumption throughout the Amazon. It is unthinkable that these biodiverse areas could disappear, yet scientists are now concerned that cloud forests could be gone in a matter of years.

Peruvians wonder how they will be able to meet impending environmental challenges, from melting glaciers to farming difficulties high atop mountain peaks to vanishing cloud forests. With uninspiring political leadership, many are placing their hopes upon the Global North to do the right thing. Unfortunately, affluent nations are not approaching the climate change issue with the seriousness that it deserves. Tensions between the Global North and South were put on vivid display at a recent climate change conference in Poznan, Poland. There, poor nations grew irate when the European Union, Canada, Australia, Japan, and Russia balked at the idea of increasing climate adaptation funds to provide environmental relief for the Global South. Needless to say, there was little progress on the issue of carbon emissions cuts, either. As poorer countries accused wealthier countries of a "vision gap," the conference collapsed and ended on a bitter note. "In the face of the unbearable human tragedy that we in the developing countries see unfolding every day, this is nothing but callousness, strategizing and obfuscation," said India's delegate at Poznan. "At this moment, I think, we can all see clearly what lies ahead at Copenhagen."[52]

Chapter Two

LIVING IN AN AGE OF EXTREMES

From the Atlantic to the Pacific coast of South America people are becoming increasingly aware of extreme weather. In particular, people are fearful of intense storms and flooding, which leave public health emergencies and large-scale human displacement in their wake. South American fears are shared worldwide, which is not surprising given that almost half of the Earth's population resides in coastal areas. Before things get better they stand to get worse: Scientists say that even if the Kyoto climate targets are reached this will not be enough to avert severe weather associated with global warming. At Poznan, Poland, during recent climate talks, experts warned that up to 250 million people could be displaced by the middle of this century as a result of extreme weather conditions and other negative impacts of global warming. In order to head off the worst effects of climate change on the entire world, the Global North must not only cut its emissions and provide adaptation funds to poor countries but also fund humanitarian aid to help environmental refugees. These climate change refugees fall into a legal limbo. While most of the displacement due to climate change is likely to occur within countries, those who do cross national frontiers don't enjoy the same protections under international law as refugees fleeing violence. True to form, affluent countries have failed to adequately address the issue of environmental displacement, let alone provide the increased material relief that will be necessary in coming decades.

Without much help in the works from the Global North, people living along the Pacific coast of Peru are bracing themselves. In particular local residents are concerned about El Niño. As long as anyone can remember in Peru, rainfall has come after a wave of hot seawater the size of Canada appears off the Pacific coast of South America. Because the ocean would warm up around Christmas time, the fishermen called the meteorological phenomenon El Niño or Christ child. The gigantic storm would pour incredible amounts of rain onto Peru's otherwise barren and arid northwestern coast. El Niño is oftentimes followed by La Niña, during which time colder-than-average water replaces warm water and the weather effects are the opposite of El Niño. Typically, periods of warm waters in the eastern Pacific (El Niño) and periods of cooler waters (La Niña) are accompanied by changes in air pressure in the east and west Pacific called the Southern Oscillation. Scientists now call the whole cycle the El Niño Southern Oscillation (ENSO).

El Niño occurs irregularly, about every two to seven years, and lasts from twelve to eighteen months. In the beginning, prevailing winds in the Pacific weaken and there's a shift in rainfall patterns. In the United States the weather phenomenon is traditionally associated with beneficial winter precipitation in the arid Southwest, less wintry weather across the North, and a decreased risk of wild fire in Florida. It was thought that El Niño helped to suppress Atlantic hurricane activity, though recent research has challenged such a view. Indeed, scientists have now linked periodic warming of the Pacific to increased hurricanes in the Atlantic.[1] Researchers say that a new form of El Niño might not only cause a greater number of hurricanes than in average years but also result in a greater possibility of hurricanes making landfall.[2]

Throughout the Pacific Basin but also farther afield, El Niño is normally linked with extreme weather like droughts and floods. While prolonged dry periods may prevail in Southeast Asia, Southern Africa, and northern Australia, wetter-than-normal weather hits much of the United States, and heavy rainfall occurs in Peru and Ecuador. Further inland in South America, El Niño has brought drought and forest fires to the Brazilian Amazon. Sadly, the regions most affected by El Niño are those with the least resources: Southern Africa, parts of South America, and Southeast Asia. For poor countries, El Niño can be economically devastating and lead to food shortages and hunger, because it gives rise to drought and billions of dollars worth of damage to agriculture. In coastal regions, El Niño can result in the collapse of fish stocks.

Could climate change alter El Niño in fundamental ways? There's been some scientific debate over the question of whether global warming has made El Niño stronger over the years.[3] In an effort to understand how El Niño works at the global level, climate experts have been intensifying their efforts. Concerned about the potential damage of future El Niños, governments have been investing in equipment designed to monitor specific conditions in the Pacific that trigger El Niño. One of the most important technological innovations of recent years has been the tropical atmosphere and ocean array of seventy moored buoys spanning the equatorial Pacific. The buoys, which were completed in 1994, are today the globe's premier early-warning system for change in the tropical ocean. Monitoring water temperature from the surface down to 1,600 feet, the buoys keep track of winds, air temperature, and relative humidity. Ingeniously, data collected by the buoys is sent to polar-orbiting satellites and then on to the National Oceanic and Atmospheric Administration's Pacific Marine Environmental Laboratory, located in Seattle.

Armed with this vital temperature data, as well as information supplied by research ships, scientists are now able to acquire a better idea of what is happening with the upper ocean and lower atmosphere. Moreover, a U.S.-French satellite measures sea-surface elevation and relays information about ocean circulation. Thanks to the buoys and the satellite, climate scientists currently have information of unprecedented range and accuracy. This has enabled them to confirm and expand their theories about what weather and sea changes signal the periodic and inevitable arrival of El Niño and La Niña. Despite these advances, some experts still say it's difficult to forecast the relative severity of future El Niños because the existing climate models don't simulate the dramatic weather phenomenon very well.[4]

Other scientists warn that we are playing with fire. Recently, an international team including some of the world's most prestigious scientific organizations warned that El Niño will become more intense and imperil sensitive ecosystems like the Amazon. The experts have become so alarmed that they have called for an early warning system to monitor such fragile ecosystems.[5] Meanwhile there's been growing scientific agreement about the increasing frequency of El Niños. The organization that sets the bar for scientific research on climate change is the U.N. Environment Program's Intergovernmental Panel on Climate Change or IPCC. But some scientists, such as Dr. Philip Fearnside of Brazil's National Institute for Research in the Amazon, say that the group has been very conservative in the way that it ratifies reports. Fearnside, himself an IPCC collaborator, says that the organization's second report in 1995 didn't address increasing frequency of El Niño events that had become statistically significant.[6]

The IPCC, however, finally came around: In a report in 2007 the group concluded that "El Niño-like conditions," meaning warm water in the Pacific, will become more frequent with continued global warming.[7] The report didn't make the specific link between El Niño conditions and droughts and floods. "But," Fearnside remarked to me during an interview in the Amazonian city of Manaus, "we know this directly just from being here [in the Amazon], you don't need a computer model to know that you have droughts and forest fires during El Niño, for example 1982, 1997, 2003."[8]

"If El Niño can be compared to a giant gun firing off climatic chaos," a reporter for the Los Angeles Times has remarked, "Peru has the geographic misfortune of being at point-blank range."[9] When Peruvians hear that El Niño might become increasingly volatile and destructive in the future it strikes fear in their midst. For thousands of years the Andean nation has had to contend with the weather phenomenon that has brought both prosperity and destruction. Archaeologists are now uncovering evidence suggesting that pre-Incan civilizations contended with harsh and severe El Niño swings.

Thousands of years ago along the Peruvian coast, pre-Hispanic ancient peoples carved out a living by hunting, fishing, and collecting mussels and clams. The strategy worked well until around 3000 B.C., when the environment began to change and the weather got bad and unpredictable. Archaeologists believe that a shift in the coupling of the atmosphere and the Pacific Ocean made El Niño more frequent. The change in temperatures killed off the local clams and mussels, making life so hard that people were forced to move inland to moon-like desert valleys. For the hunters and gatherers who abandoned coastal life, subsistence a scant ten miles away was hard. The settlers had to learn how to cultivate crops and irrigate them from the precious few nearby rivers and streams. El Niño storms would bring them water but also terrifying destruction.

El Niño continued to batter Peru, wiping out the Moche and Lambayeque pre-Hispanic civilizations a thousand years ago. The weather phenomenon may have also struck fear into the hearts of the Chimú people who inhabited the ancient city of Chan Chan. Recently, archaeologists excavating at the site uncovered the skeletal remains of a woman dating from the mid-fifteenth century. Experts believe she was buried alive, possibly as a sacrificial offering to ward off El Niño. The Inca who followed on the heels of the Chimú were clearly aware of El Niño: They constructed their cities on the tops of hills and kept stores of food in the mountains. If the Inca did build on the coast, it was not near rivers. As a result, many of their dwellings are still standing today.

Since ancient times, El Niño has also exerted a profound effect on Peruvian fisheries. Anchoveta, a type of anchovy that frequents the Peruvian coast out to eighty miles from shore, is a small fish measuring not more than twenty centimeters long. It is silver with a large mouth and pointed snout. Anchoveta are filter feeders, form large shoals, and represent a valuable source of food as well as oil. Both anchovy and sardines have served as staples for coastal peoples. At Cerro Azul, a site dating to just before the Inca conquest in 1470 A.D., people lived off anchovy and sardine and then shipped the fish to surrounding towns. The small fish were dried, stored, and in time transported inland via llama caravan. Because of their small size, anchovy and sardine can be spread out on a pavement of beach cobbles to dry in a single day. At Cerro Azul, twelve noble families commanded retinues of retainers and oversaw hundreds of anchovy and sardine fishermen. The fishing folk also captured twenty other larger species of fish, including mullet, flounder, sea catfish, and even small sharks and rays. For local residents, fish carried important cultural meaning: Burial sites included whole fish as food for the afterlife.

Over the centuries, sardine and anchovy have switched off as more common in the local diet. Though there are times when both anchovy and sardine are present, each species has a different life cycle and requires different temperature and nutrient conditions. These fish in turn depend on plankton, consisting of microscopic plants (phytoplankton) and animals (zooplankton). Microscopic plant life that floats in lighted surface waters, phytoplankton forms the base of the marine food web. With the help of the green pigment chlorophyll, phytoplankton uses sunlight to convert nutrients into plant material. While anchovy feed on large zooplankton, sardines consume phytoplankton as well as small zooplankton.

Researchers have suggested that the Pacific Ocean climate shifts back and forth between a "sardine regime" and an "anchovy regime." Anchovies thrive in colder waters and in general it was anchovy that predominated in the ancient diet. Then, in 1500, just before the Spanish arrived in Peru, the regime shifted back to sardine, corresponding to a warmer phase of El Niño.[10] Without the intellectual contribution of Alexander von Humboldt, a scientific explorer, we might not yet understand the inner workings of fish and the climate regime on the Peruvian coast. Humboldt studied the cold upwelling along the Pacific coast of South America, known as the Peru Current, and laid the cornerstones of modern-day climatology and meteorology.

Today, the Peru Current is also known as the Humboldt Current in honor of the European explorer. Flowing north to Peru from the chilly southern waters off Chile, the Humboldt Current is thought to be the planet's most productive marine ecosystem, in large measure because cold, deep waters high in nutrients interact with the rays of the sun to produce life. According to scientists, the impact of the Pacific upwelling is significant: Nutrients rise to the sunny upper ocean layers, thereby supporting the tiny plankton that feed on them. This food chain progresses to anchovies and sardines, which in turn support many larger fish. Coastal upwelling regions, including Peru, northern California, Oregon, and the west coast of Africa, contain some of the world's richest fisheries. Today the Peruvian coast has one of the highest fish densities in the world and the planet gets a full one-fifth of its fish stocks from the southeastern Pacific area.

During El Niño, however, the cold waters of the Humboldt Current are displaced by warmer water coming from the central Pacific. The displacement gives rise to a reduction in plankton, the key building block of the Humboldt food chain. This in turn creates food shortages all across the area. Along coastal Peru, anchoveta becomes scarce and some fish, such as hake, a large bottom fish, migrate down the continental slope into colder, more suitable ocean depths. A marine fish resembling cod, hake has two dorsal fins and an elongated body and is considered commercially valuable as a food source.

Though some of the world's upwelling areas have demonstrated remarkable resilience through El Niño cycles, others have not. Could global warming push marine ecosystems just a little too far? That has been the precise concern in Peru in recent decades as the country has been pummeled by fierce El Niños. During the 1972–73 El Niño for example, Peruvian fisheries were decimated. As a result of the meteorological disturbance, up to 1,500 fishing boats and 200 fish processing plants went out of business and the anchoveta catch went down by about seven million metric tons. Meanwhile, more than 100,000 people linked to the fishing industry were thrown out of work.[11]

If the implications for fish species were not serious enough, other wildlife may face difficult challenges ahead in an era of global warming. Peruvians are already becoming accustomed to calamitous ecological collapses like those that have accompanied recent El Niños. Take for example the case of the 1982 El Niño: In May of that year the easterly (east to west) surface winds that usually extended almost the entire way across the equatorial Pacific from the Galapagos Islands to Indonesia began to decrease in strength. To the west of the dateline winds shifted to a westerly direction, ushering in a period of stormy weather. Over the course of the next few weeks, the ocean started to react to changes in wind speed and direction.

At Christmas Island in the mid-Pacific, the sea level rose by several inches. Farther to the west, typhoons were driven off course to islands like Hawaii and Tahiti. Meanwhile, monsoon rains fell over the Central Pacific instead of the Western Pacific, prompting drought and terrible forest fires in Indonesia and Australia. The weather took an ominous turn by October, when sea levels rose up to a foot in Ecuador, some 6,000 miles to the east. At the Galapagos Islands, which belong to Ecuador, sea surface temperatures rose from the typical low 70s Fahrenheit to well into the 80s. In the United States, widespread flooding occurred across the South, while owners of ski lodges in the north reported unusually mild weather and a lack of snow. In California's San Mateo County, heavy surf and rains dramatically eroded beaches and fragile sea cliffs. Landslides meanwhile damaged the coastline and the main highway was closed. Even worse, buildings were damaged and put at risk by the retreat of sea cliffs and slope failure.

When above-normal temperatures along the entire Pacific coast stretching from Chile to British Columbia prompted tropical and subtropical fish to migrate or be displaced toward the poles, Peruvian marine wildlife suffered: 25 percent of the fur seal and sea lion adults died off as well as all of the pups.[12] Seal pups are in danger during El Niño, probably because their parents are unable to find food. Seals feed on anchovies and sardines and they normally make quick, shallow dives for fish at night, when schools are close to the surface. But when El Niño drives warm water toward the Peruvian coast that warmer water forms a layer over the cold, nutrient-rich water that supports fish. As a result, fish schools are driven down to depths that the seals can't reach. In a mild El Niño, cold waters persist longer and there's still food around on the coast. But scientists say that if super strong El Niños occur with greater frequency this could knock out the local seal population.

In an era of potentially more severe El Niños experts are also worried about seabirds. Desperate for food during the 1982 El Niño, they abandoned their young and fanned out over the wide Pacific Ocean expanse. In the future scientists will be careful to monitor the blue-faced booby and the great frigate. A bird that got its name from its tameness and lack of fear of humans, the booby allowed early mariners to easily hunt it down. The booby has a wingspan of almost five feet, dives for small fish and squid, and accomplishes this feat from one hundred feet above the surface. It forages up to fifty nautical miles from land. Frigate birds have hooked bills and are excellent gliders. They do not land in the water, preferring instead to snatch jellyfish and marine creatures from the ocean surface or steal food from others. Known as mischievous aerial pirates, frigates are wont to harass boobies, gulls, or terns until rival birds drop their fish. Frigates have astounded scientists with their endurance and long-range flight.

In light of global warming, what is to become of the booby and frigate? Scientists worry that El Niño adversely affects the growth and reproductive success of seabirds. During El Niños, the booby and frigate are forced to abandon their nestlings, which results in mass starvation of young. Even adult birds perish of starvation if El Niño is prolonged. Yet another emblematic species, the Humboldt penguin, makes its home on Peruvian shores. The Humboldt eats small fish like anchovies and sardines. However, now that its food supply has come under stress as a result of El Niño, the penguin is in jeopardy and its numbers have suffered.

What kind of effect might El Niño have upon human beings? If the dramatic meteorological events of 1991 are any indication, Peru could be in for a rocky road. In that year, El Niño led to a major cholera outbreak in the Andean nation. El Niño is linked to the disease because warming ocean currents encourage the growth and spread of bacteria. Scientists have shown that bacteria that cause cholera often live on microscopic plants and animals, called plankton, that drift in the ocean. Because the organisms are sensitive to heat, warmer waters promote their growth and increase the risk of cholera outbreak.

What probably happened in 1991 was that fish and shellfish fed on an increased and infected population of plankton. That spelled problems further up the food chain, as Peruvians are fond of a seafood dish called ceviche. The specialty is prepared by marinating raw fish in lemon juice. People who consumed infected seafood became ill, and poor hygiene spread the infection to the rest of the population. At the time, Peruvian scientists remarked that people simply panicked. It was the first time that cholera had struck the Americas in the twentieth century and the Peruvian people had little natural resistance to the disease. Because of the widespread lack of clean drinking water and adequate sewage disposal, the population was particularly vulnerable.

In 1991, experts believe that a whopping 14 million Peruvians got infected with cholera and a full 350,000 wound up in the hospital.[13] The outbreak, moreover, proved economically devastating for Peru, costing the country about $770 million.[14] With cholera, many people can be "healthy carriers" and be infected but show no symptoms. Cholera is a very nasty disease; it feels like you've opened a faucet in your system. Water simply comes out of your body in large amounts. The tragic result of the cholera epidemic was evident in Lima's state-run Dos de Mayo Hospital. There, admissions hit 162 a day and the list of patients crowded off the upper edge of a blackboard chart. One male patient remarked "I bought a cup of herbal tea from a street vendor, and an hour later I was throwing up." As he spoke from his hospital bed, an overhead serum bottle dripped dextrose solution into his right arm.

The patient lay in a former women's ward that had only just been re-labeled "Cholera II" by a hurried member of the hospital staff. The man's wife remarked to the *New York Times* that two people on her block had died of cholera, and that within their shantytown neighborhood, people drew water from a communal well. The capital city of Lima, which had expanded from one million people in 1950 to seven million by the early 1990s, was ill prepared to deal with the crisis. Already facing terrorism, hyperinflation, and foreign debt obligations, the government could barely maintain existing water and sewer systems, much less expand the network of pipes to half of the city that did without.

Though the 1991 El Niño certainly took a tragic human toll in Peru, experts praised doctors for keeping the death rate below 1 percent.[15] Because there were sufficient health clinics and people got the message to come in fast, the death rate was relatively low at 3,500 people. Cholera is easily treated, given availability of clean water for rehydration and essential salts to replace any lost body salts. Unfortunately, in remote areas of Peru many people live ten hours from adequately supplied health clinics. Cholera can kill a person in exactly that amount of time. Some public health officials criticized the country's politicians for seeking to minimize the epidemic in order to reduce its impact on Peru's food and tourism industries. "The public information campaign was a disaster," remarked Peru's former health minister.[16]

Scientists also fear that climate change might be fueling dengue fever. A mosquito-borne infection, dengue symptoms include fever up to 105 degrees Fahrenheit, terrible headache, eye pain, joint and muscle pain, nausea and vomiting, and a rash appearing over most of the body. Dengue hemorrhagic fever is even more serious and includes all of the classic symptoms of dengue plus marked damage to blood and lymph vessels and bleeding from the nose, gums, or under the skin, which results in purple bruises. An even more extreme disease, dengue shock syndrome, is sometimes fatal and usually occurs in children and young adults. Though the exact connection between dengue and weather conditions is unclear, it appears that higher temperatures tied to El Niño might have an effect on virus transmission. Yet another worry is diarrhoeal disease, which is the second leading infectious cause of childhood mortality worldwide, accounting for a total of around 1.8 million deaths annually. An illness that is sensitive to climate change, diarrhoeal disease in the Peruvian capital of Lima is three to four times higher in the summer than in the winter and goes up by 8 percent for every 1 degree Celsius increase in temperature.[17]

Six years after the 1991 El Niño, Peru was once again struck with a public health emergency. Rising out of the Pacific, another El Niño bore down on the coast. According to *National Geographic,* by the time it had run its course El Niño altered weather patterns around the world, killed about 2,100 people, and caused at least $33 billion in property damage.[18] Globally, El Niño led to sometimes-bizarre weather conditions: In the United States the winter turned balmy in the otherwise frigid U.S. Midwest. Indonesia and Southeast Asia suffered from drought and wildfires. In Africa, a major drought hit the continent.

Across the border from Peru in neighboring Colombia, two million chickens dropped dead after a heat wave linked to El Niño struck the country.[19] To make matters worse, the drought resulted in forest fires. Elsewhere in the Andean region, the Ecuadoran government estimated that it would cost $2 billion to repair roads, bridges, and infrastructure damaged as a result of El Niño. In one freak incident, a landslide caused by El Niño-driven rains cracked the country's largest oil pipeline, sparking an explosion and fire that killed eleven people, burned eighty and dumped 8,000 barrels of crude oil into a local river. In Venezuela, drought obliged hydroelectric plants to ration power. Brazil meanwhile witnessed its worst drought in twenty-five years, which accelerated fires that spread to virgin areas of the Amazon rainforest. In Guyana, the economy teetered on the verge of disaster as drought dried up rivers and streams, forcing hundreds of miners to abandon remote gold mines in the jungle. According to mining officials, gold accounted for 40 percent of the country's exports.[20]

The El Niño of 1997–98 resulted in a public health scare across South America. In Argentina, concerns mounted that a plague of mosquitoes might cause an outbreak of dengue fever. In the capital of Buenos Aires, high-risk areas were fumigated to stop mosquitoes from spreading. In something out of a strange sci-fi movie, panicky Chileans were shocked by the emergence of giant killer rats. Startled, the authorities reported that they were unable to deal with the rodents.[21]

In Peru, it fell to President Alberto Fujimori to reassure the panicked population. An engineer, mathematician, and former university professor, Fujimori had built up his political career by presenting himself as a can-do, pragmatic technocrat.

Apparently under pressure from the country's powerful fishing industry, he as well as his wife, Susana, and the ministers of fisheries and agriculture appeared on television eating ceviche. Lounging at a beach-front restaurant, the president sat down to sashimi while surprised foreign correspondents looked on. Fujimori's public relations move seemed to backfire later, however. Within days, cholera hospital admissions soared. Even more embarrassing, one of the reported patients was the government's own fisheries minister, Félix Canal. It took Canal a week to recuperate in a military hospital. Through a spokesman, the minister denied that he had fallen sick with cholera, instead claiming that he had fallen ill with a bad case of laryngitis. Meanwhile, the disease spread from the coast to the Andes and from there on to the Amazon and the jungle town of Iquitos. Cases were also reported in neighboring Ecuador, Colombia, and Brazil.[22]

Every time South Americans are hit with a powerful El Niño and a cholera or dengue outbreak, they wonder what it's going to take for the Global North to spring into action. By 2030 many poor countries could be spending billions of dollars to manage additional costs to health services as a result of climate change. What's more, poorer countries stand to be disproportionately affected by public health disasters because vector-borne diseases are more common in tropical nations than in many affluent countries. Indeed, the Global South is more vulnerable to such outbreaks due to low socioeconomic development and poor coverage of health services. This means that the per capita mortality rate from vector-borne diseases is nearly 300 times higher in poor countries than in developed regions. With so much at stake, richer nations must contribute more funding to much-needed adaptation funds so that the Global South is not left to face the ravages of climate change on its own.[23]

Though climate change-spurred public health emergencies are certainly a huge concern, there's another looming threat that the planet must deal with: human displacement and environmental refugees. According to scientists, climate change will result in dramatic movements of people across the world in the twenty-first century. From droughts to water shortages to floods to storm surges to sea-level rise, the effects of global warming stand to overwhelm ill-prepared poverty-stricken nations. Today, environmental refugees are already outnumbering refugees escaping from war, and their numbers will reach into the millions as they flee their homelands. Africa, coastal areas of Asia, and the Pacific Islands will be most threatened by the environmental refugee problem, though Latin America is clearly already feeling the pinch.

In Peru, it is El Niño-driven floods that most strike fear into the hearts of people living along the coast. During the catastrophic El Niño of 1997–98, Peru suffered over ten times more rainfall than usual. Few imagined that El Niño could dump so much rain on the coast, in some places up to five or six inches a day. In the hamlet of Chato Chico in the north of the country, farmers had to endure weeks of incessant rain. On top of that, the nearby Piura River simply would not stop rising. At long last the river simply overspilled, pouring water into Chato Chico. For the town's residents, the event was terrifying. As the river overwhelmed their houses, farmers found themselves chest-high in water. Suddenly they were surrounded from all directions. The flooding dragged away all the small animals, and then the houses simply fell down completely.

Frantically splashing through the muddy flood, hundreds of families tried to save whatever they could. They grabbed clothes for the children and everything else—including chickens, goats, pots, pans, religious icons, as well as personal treasures—just washed away. Fortunately some residents were evacuated on barges and helicopters to a dry refugee camp in the desert and nearly all survived. Every day, farmers would walk three miles each way to farm riverside fields that lay right next to their former village. "We can't go back," one refugee remarked in a resigned tone of voice. "It will happen again. If God wants to save us next time too, we say thanks. But right now, this is where we will stay." Though those farmers survived the flood, other Peruvians were not so fortunate. Sixty miles south of Chato Chico a small village called Motse outside the city of Chiclayo was submerged in minutes, drowning ten people.[24] By the time it was all over, El Niño had inundated whole cities by flooding and landslides, and 200 people lay dead.[25] The torrential rains were the worst flooding in fifty years and left tens of thousands homeless.

How will cash-strapped governments in the Global South cope with El Niño and the specter of increasing numbers of environmental refugees? During the 1997–98 El Niño, the task of dealing with coastal flooding fell to President Alberto Fujimori. Dressed for action in boots, jeans, and a windbreaker, Fujimori hit the ground running in the flooded southwestern town of Ica. Striding through surprised crowds at a market under reconstruction, he promised incensed housewives he would restore water service and deployed his ministers to oversee recovery tasks, all the while exhibiting an impressive command of rainfall statistics, street names, and other minute details. Flying around the Peruvian coast to survey the destruction, Fujimori displayed a singular leadership style that thrived on crisis.

Fujimori's critics claimed that he was exploiting El Niño to advance a third election bid in 2000. The president's supporters, however, countered that Fujimori had always traveled across the nation and behaved in this manner. Seeking to cultivate a down-home image, the president came to the rescue of the people with rolled-up sleeves. In a televised incident, he spotted a group of stranded flood victims from his helicopter. Ordering the pilot to touch down, Fujimori hoisted the women and children aboard to safety. In the country's third largest city, Trujillo, the president personally oversaw the unloading from his plane of crates that contained one hundred collapsible potable water containers. Climbing behind the wheel of a red SUV, he sped away from the airport. People standing in dilapidated streets did double takes and shouted familiar greetings: "Fuji!" "Alberto!" and "Chino!" (the Spanish word for Chinese, a common term used in Peru even for Peruvians of Japanese descent like the president). In a surreal moment Fujimori visited a slum where waters had flooded an adjacent cemetery, emptying 123 graves. The people of Trujillo were subjected to a horrifying apparition: their own dead relatives floating down local streets in the very same Sunday attire in which they had gone to the grave.

Despite the human toll of El Niño, many Peruvians said that the destruction would have been worse without the government's disaster relief action. Prior to the environmental tragedy, Fujimori carried out a $150 million infrastructure and preparedness program that revamped the drainage systems in cities like Piura and Tumbes in the north. Then, after disaster struck, Fujimori again sprung into action. In formerly desert regions that had become bizarrely lush, Fujimori planted trees and built irrigation systems in an effort to reclaim once barren regions for cattle ranching. In the short term, these efforts increased his popularity.[26] But in the long term, problems with delivering relief played into the hands of the opposition. When the president's cabinet estimated nationwide damage at $800 million, angry opposition politicians shot back that the damage was closer to $1.8 billion. The current aid levels were unsatisfactory, they claimed, and Fujimori's relief efforts amounted to nothing more than adventure tourism.[27] Fujimori's standing fell further as the country underwent recession as a result of El Niño and major flood damage. [28]

Though El Niño has wreaked havoc upon the Pacific and Peru, another specter now looms on South America's Atlantic flank. In 2004, shortly before Hurricane Katrina hit New Orleans, a truly bizarre storm hit Brazil. In mid-March, a nameless low developed far off the Brazilian coast. Later, however, the system took on tropical characteristics, such as a warm core with little temperature contrast. Already, Brazilians in the south of the country had been left scratching their heads at the peculiar weather: January and February were the coldest they had been in twenty-five years. Indeed, the storm formed over waters that were actually somewhat cooler than average. At the same time, the air was much colder than normal, which produced the same type of intense upward heat flux that fuels hurricanes in warmer waters such as the Caribbean.

"Before long," notes the *Guardian* of London, "the heat flux and light shear gave birth to a system that bore the satellite earmarks of a hurricane."[29] As scientists watched the storm approach Brazil, they grew puzzled. Pick up most any textbook on hurricanes, and it will tell you that the one spot where hurricanes do not occur is the South Atlantic. According to conventional wisdom, atmospheric conditions in the region do not provide enough spin near the surface to even get hurricanes started. What's more, winds high in the atmosphere tend to shear off storms.

The historical record is replete with cyclones that have plagued the North Atlantic and wreaked havoc upon shipping and the like. In 1564 the French decided to challenge Spanish preeminence in the New World. They built a fort near present day Jacksonville, Florida, and pursued a Spanish fleet down the Atlantic coast the following year. At Saint Augustine, however, both fleets encountered a terrible storm and the French were shipwrecked. As a result, the French lost their bid to control the Atlantic coast of North America. Yet another devastating hurricane hit five ships belonging to the Virginia Company in 1609. The ships were dispersed by the storm and shipwrecked on Bermuda. The hurricane inspired Shakespeare to write the *Tempest,* first performed in London in 1611. Much of the play's language recalls the words of the hurricane's survivors.

The rub, however, is that these hurricanes occurred in the North Atlantic and not in the south. Indeed, no hurricane had ever been reported in this geographical area. As a result, Brazilian fishermen were totally surprised and caught off guard when Catarina struck and engulfed their boats. The storm was quickly dubbed Hurricane Catarina and developed an eye and other well-defined features. When scientists talk about an eye, they are referring to a roughly circular area characterized by comparatively light winds and fair weather at the center of a severe tropical cyclone. The calm eye of the storm shares many characteristics with tornadoes, waterspouts, dust devils, and whirlpools. At the eye's axis of rotation the winds remain calm but strong winds may extend right inside the eye itself. There is little or no precipitation within the eye, and the region has the lowest surface pressure and warmest pressure aloft. Eyes may vary in size, anywhere between 5 miles across to a monster 120 miles. The well-known swirling around the eye is actually a low-pressure vortex that forces high pressure downward. There, it is warmed by the sea and fuels storm clouds that give rise to driving rain.

As they poured over the satellite data, researchers may have wondered whether Catarina itself could become one of these monsters. Acting prudently, they alerted the Brazilian navy, which was responsible for analyzing weather forecasts. When it saw what was coming, the navy in turn grew alarmed and notified the civil defense unit and regional weather center of the Brazilian state of Santa Catarina. By this time, the storm was bearing down on the Brazilian coast as a Category 2 hurricane. As the 85 miles per hour storm came aground on March 28, 2004, some 500 miles southwest of Rio de Janeiro, some local residents fled while others tried to ride out Catarina. Many Brazilians were unprepared for Catarina's ferocity and had not constructed their houses to withstand high winds.

Though Catarina weakened somewhat to a Category 1 hurricane, the storm nevertheless wound up killing eleven people and left a trail of destruction throughout Santa Catarina and the state of Rio Grande do Sul. In Santa Catarina alone, the hurricane damaged more than 32,000 houses and destroyed more than 400.[30] In addition, many cities and towns had their water and drinking supply cut off. The storm moreover damaged public buildings and destroyed a hospital. Catarina caused catastrophic economic damage, amounting to more than $340 million, while crop losses were estimated at $34 million.[31] In the city of Torres, the local fisher folk stayed inside their houses and trusted in God as the sea rose. After the wind came the rain, and in the morning people were greeted by damaged houses, broken windows, furniture on the ground, and fallen trees.[32]

In the wake of the storm, debate swirled amongst scientists as to the true nature of Catarina and why it had developed in an area where there had previously been no hurricane activity. Typically, hurricanes start out as a cluster of ordinary thunderstorms. Strengthened by heat from warm tropical waters the storms then swirl together and join to form first a tropical depression, then a tropical storm, and finally a hurricane. Hurricanes forming over warm waters gain strength from the heat, which is released as evaporated water from the sea surface, and condenses and falls back to Earth in the form of intense tropical rain. The resulting heat that is released drives rapid updrafts that cause more water to evaporate from the ocean's surface.

This in turn leads to a self-reinforcing vortex of swirling clouds that can give rise to wind speeds of more than 100 miles per hour. Catarina had the same familiar swirl of clouds typical of hurricanes, in addition to a well-defined eye. A team of observers reported strong winds, heavy rainfall, and cold temperatures, followed by calm winds, rapid warming, and a clear sky before Catarina made landfall. Afterward, there was a second bout of heavy rain, cold temperatures, and swirling winds. While

Catarina had the warm core and clear, sharp eye of a hurricane, the height of its clouds fell short of hurricane standards. It's unusual, note experts, that Catarina developed an open eye, warm core, and banded wind structure given the low ocean temperatures.[33]

Was Catarina simply a rare event at the extreme edge of the normal bell curve of South Atlantic weather—or was it a kind of "threshold" event, foreshadowing an ominous and abrupt change in the planet's climate system? Scientific discussions of environmental change and global warming are haunted by the specter of nonlinearity. Climate models are easiest to build and understand when they are simply extracted in a linear way from well-quantified past behavior. On the other hand, all the major components of global climate, including air, water, ice, and vegetation, are nonlinear. When they reach certain thresholds, these components may switch from one regime to another, with terrible consequences for species that are too finely-adapted to the old regime. Scientists used to believe that major climate transitions typically took place over the course of centuries or even millennia. Now, however, as a result of our study of ice cores and sea-bottom sediments, we know that global temperatures and ocean circulation can change abruptly under certain conditions, in a decade or even less time.[34]

Though Catarina was the first documented hurricane in the southern Atlantic, researchers believe it may not be the last. Indeed, the signs that we are seeing in the south Atlantic are certainly consistent with an enhanced occurrence of such extreme weather. Such signs include very strong evidence that the increase in greenhouse gases and depletion of the ozone layer are related to decreased atmospheric pressure off the Antarctic coast and a rise in pressure further north. That in turn has had an impact upon the southern tropics, where wind shear (the difference in wind speed at high and low altitudes) is dropping. Catarina, scientists say, was like the little sister of hurricanes Katrina or Rita.[35]

From the Pacific to the Atlantic, South America has been pummeled by extreme weather. While Peruvians would surely like to see a slowdown in

the ravages of El Niño, there's no sign that the meteorological phenomenon is letting up. Indeed, in 2009 scientists warned of wild weather as
El Niño exacerbated global warming. Development groups meanwhile
say the arrival of a new El Niño in 2009–10 is worrying as it could add
to the effects of climate change, worldwide economic recession, and
hunger. What's more, experts warn that the next few years will be the
hottest on record and will feature droughts, floods, and other extreme
events.

With so much at stake, the Global South has asked affluent nations
to step up to the plate. Poor countries that stand to be affected by extreme coastal weather have called on the affluent north to accept climate
refugees. Experts have even lobbied for a new legal instrument specifically
suited to the needs of climate refugees—a Protocol on the Recognition,
Protection, and Resettlement of Climate Refugees under the United Nations Framework Convention on Climate Change. If the Global North
is forced to recognize the urgent issue of environmental refugees then
perhaps this could lead to progress in other areas. Specifically, poorer
countries want more aid in the form of adaptation funds—and a lot of it.
Experts estimate that poor countries need between $5 and $10 billion to
help deal with climate change and severe weather. By 2020, that figure
could increase to a whopping $100 billion every year.

During recent climate negotiations in Poznan, Poland, the Polish environment minister warned of future perils. Without concerted action,
he said, the world would confront more intense flooding, cyclones of
more destructive power, and pandemics of tropical disease. But in the
midst of an economic slowdown, industrialized nations at the conference
were in no mood to pay any attention, let alone put down increased financial resources to confront climate change. Undeterred, the minister
pleaded with his colleagues. "Do not let particular interests change the ul-
timate goal," he said, adding "we have to change the development path
taken by humanity."[36]

Chapter Three

CARBON CONUNDRUM

While Peru and Brazil will surely have difficulty confronting calamitous weather in the years to come, some in the Global North might wonder "What does it all have to do with me?" Unfortunately what happens in Peru and Brazil, the two largest Amazonian countries, has a big impact on global climate. Indeed, the same one-two punch from the Atlantic and the Pacific that has brought so much death and destruction to the two countries now stands to unleash drought in the rainforest—drought which stands to affect us all. Saving the Amazon rainforest became something of a cause célèbre in the 1990s, when many Americans became aware of the plight of indigenous peoples and the vast deforestation in Brazil and neighboring countries. What was missing from the debate, however, was the essential role played by the rainforest within the world's climate system.

The prospect of the Earth's rainforests turning into a collection of tinder boxes is of huge concern to scientists, and with good reason: The forest releases vital oxygen and removes large quantities of carbon dioxide, a pollutant harmful to humans, from the Earth's atmosphere. Carbon dioxide is probably the most important of the greenhouse gases. When we burn fossil fuels, either by running our vehicles, heating our homes, or powering our factories, we release it. Fortunately, the Amazon acts as the planet's air conditioner by sucking up millions of tons of greenhouse gases and storing them safely out of the atmosphere. In the rainforest, most of this carbon gets locked up in living vegetation, dead wood, and decaying leaves.

Sounds good, but the Amazon can also be a curse: When there are fewer trees there is more carbon dioxide. It's a downward spiral that has become all too familiar in recent years: As forest cover is removed, more severe weather patterns emerge, with hail and lightning igniting fires. According to scientists, global warming has led to an increase in electrical storms, especially in the tropics. In Brazil there's been an upswing of lightning storm activity that has resulted in human deaths as well as great economic losses. The smoke from fires in turn inhibits further rainfall. When there's large-scale tree burning, smoke escapes into the atmosphere where it adds to the carbon dioxide blanket over the Earth. Carbon dioxide accumulates in the atmosphere, stopping heat from the sun from escaping the Earth, thus adding to global warming.

The Amazon contains about one-tenth of the total carbon stored in land ecosystems and recycles a large fraction of rainfall that falls upon it. But now, deforestation in the tropics is releasing about 1.5 billion metric tons of carbon into the atmosphere every year.[1] Deforestation should not be underestimated in terms of its long-term ecological effects: Currently, rainforest destruction gives rise to 20 percent of greenhouse gas emissions. Were the Amazon rainforest to collapse, 26 billion tons of carbon would be released into the atmosphere, about as much as the entire planet produces over the course of two and a half years.[2] Ultimately, then, climate change may wind up transforming Amazonian forests from a net sink or absorber of atmospheric carbon dioxide into a source. This will further contribute to dangerous levels of atmospheric carbon dioxide. Brazil and Peru, two countries that already have their fill of environmental problems as a result of climate change, could be stretched to maximum capacity as they seek to cope with global warming and the Amazon.

Interestingly, however, the rainforest might yield the key to our carbon conundrum. Today scientists have become very interested in *terra preta de índio*, Portuguese for "Indian black earth." A unique, mineral-rich soil, *terra preta* was created by pre-Columbian people by adding charcoal and animal bones to regular soil in order to create a highly fertile hybrid ideally suited for agriculture. In contrast to today's Amazonian farmers who practice slash-and-burn agriculture, pre-Columbian farmers carried out the so-called "slash-and-char" method. Instead of completely burning trees to ash, indigenous peoples smoldered organic matter to form charcoal, which they simply stirred into the soil. One benefit of the technique was that it released less carbon into the air than today's slash-and-burn method. Unfortunately, slash and char was widely abandoned over the past 500 years: When the Europeans arrived, they destroyed ancient Amazonian civilizations and those Indians who survived disease and warfare were forced deeper into the jungle, where they became hunter gatherers.

Scientists have become fascinated by *terra preta* because it can pull large amounts of carbon out of increasing levels of carbon dioxide in the Earth's atmosphere and thus help to prevent global warming. Researchers hope that *terra preta* can help to restore degraded soils, increase crop yields, and lead to technologies that will allow us to sequester carbon in the earth, thus preventing critical changes in world climate. It is estimated that up to 10 percent of the Amazon basin may contain the soil—currently *terra preta* is excavated and sold as potting soil. If archaeologists are correct, *terra preta* not only removed large quantities of carbon from the atmosphere but may continue to exert a beneficial environmental impact since *terra preta* charcoal can retain carbon in the earth for almost 50,000 years.

In ancient times, *terra preta* also supported literally millions of people. The rainforest has a long history of human settlement and, contrary to popular belief, large and complex societies existed in the region prior to the arrival of the Europeans. Indigenous peoples planted trees in fertile river basins and created orchards, which allowed the natives to survive during periods of drought. It's thought that the Indians were able to support large populations through careful management of the environment, unlike the Europeans, who imposed single-crop monocultures. Consider: Through *terra preta* the Indians were able to sustain a large settlement called Marajó that may have supported more than 100,000 people.

Anna Roosevelt has played a key role in excavating the site. An anthropology professor at the University of Illinois at Chicago, she's a veteran of more than fifty archaeological expeditions. She is hardly the first member of her family to be attracted to the Amazon: In 1913, her great-grandfather Theodore Roosevelt joined an expedition to map an uncharted Brazilian river named Rio da Dœvida or "River of Doubt" (the river was subsequently renamed "Rio Roosevelt"). During its travels, the party was beset with all kinds of troubles including accidents, disease, and even mutiny. Roosevelt himself became so sick at one point that he begged to be left to perish rather than be a burden to the expedition. Although Roosevelt managed to return home, his death five years later was linked to health problems contracted during his Amazon misadventure. When she's compared to Teddy, Anna Roosevelt bristles. "He wasn't a real explorer," she remarks. "He did two little explorations and I've done 50. No one would mistake what he did for research or exploration." Recently, Roosevelt was asked to participate in a reenactment of her great-grandfather's expedition. She categorically refused. "The interest in Theodore Roosevelt competes with interest in me," she declared.

Anna Roosevelt's concern about being eclipsed is slightly overstated. Indeed, through her work Roosevelt has challenged prevailing dogma and is helping to rewrite the history of humankind in the Americas. Specifically, her work at Marajó has confounded stereotypes and drawn attention to the issue of *terra preta*. Located on a barren island the size of Switzerland at the mouth of the Amazon, Marajó recently served as the site for the Brazilian version of the hit TV show *Survivor*. Roosevelt, who eats grilled piranhas out in the field during her archaeological expeditions, found hundreds of large earthworks on the forest floor, each about sixty feet tall and covering up to a square kilometer. Remains inside the 1,800-year-old mounds demonstrated that they had been the focal points for numerous urban centers. Roosevelt concluded that the entire island had been full of large towns and was crisscrossed by roads, irrigation, and drainage networks. She called the site "one of the outstanding indigenous cultural achievements of the New World."[3]

Even more tantalizing, Marajó contained *terra preta*, and the cultivation of the soil seems to have enriched the surrounding environment rather than caused the typical ecological degradation that one would normally expect to find at such a large archaeological site.[4] What's more, Marajó is just the tip of the iceberg: All over the Amazon *terra preta* can be found at large archaeological sites inhabited by indigenous peoples. For years, archaeologists figured that the Amazon could not have supported millions of people and that Indians must have lived in semi-nomadic bands. How to reconcile such an idea, however, with eyewitness accounts from the colonial era? In 1542, Spanish explorer Francisco Orellana went up the Amazon River looking for the fabled gold city of El Dorado. When he entered the Rio Negro, a large Amazon tributary, he wrote: "There was one town that stretched for 15 miles without any space from house to house, which was a marvelous thing to behold. There were many roads here that entered into the interior of the land, very fine highways. Inland from the river to a distance of six miles more or less, there could be seen some very large cities that glistened in white and besides this, the land is as fertile and as normal in appearance as our Spain."[5]

Could Orellana have embellished his tale or flat out lied? For years scientists dismissed the account. As productive as the rainforest may seem, they argued, the native soil was unsuitable for farming or sustaining large populations. Now, however, experts are reconsidering Orellana. Indeed, archaeologists have demonstrated that the distribution of *terra preta* correlates favorably with the places Orellana reported on in the sixteenth century, an immense area twice the size of the United Kingdom. *Terra preta* is pervasive along the banks of the Amazon River and its key tributaries such as the Rio Negro and the Tapajós. Though it's unclear whether the special soil extends further inland from these rivers, some researchers believe that *terra preta* will eventually be found deep in the heart of the Amazon and could comprise up to 10 percent of total rainforest earth.[6]

If ancient cultures were able to minimize environmental damage to the forest through skillful use of *terra preta*, we are surely living in an ecological dystopia now. In contrast to pre-Columbian Indians, we have increased our carbon conundrum through deforestation and callous industrial emissions. When scientists talk about carbon, the discussion has an abstract ring to it, but make no mistake: Deforestation and elevated carbon levels in the atmosphere have real consequences. Indeed, researchers anticipate that the rainforest will become hotter and drier over time. As they dry out, forests will not be able to put water vapor into the air, which will create a cycle of less and less precipitation.

Such a prospect hardly provides a desirable environmental scenario for the future. Normally rainforests act as precipitation-inducing machines. Approximately half of all Amazonian rainfall is almost immediately restored to the atmosphere as water vapor via plant respiration. This in turn helps to sustain cloud cover and yields frequent rainfall, particularly in the dry season when forests are susceptible to drought and fire. While some argue that forests may grow faster under a global warming regime, others claim they might be more likely to burn, suffer from disease, or die out from drought. One thing's for sure, though: Higher levels of atmospheric carbon dioxide are fundamentally altering the nature of the Amazon jungle. Indeed, rainforest trees are now growing more rapidly, though they are also dying faster.[7]

How will forests react in the long term to such changing environmental and climatic conditions? In light of the natural history, some researchers doubt that the Amazon is on the verge of a major "dieback" resulting from climate change.[8] According to the so-called "dieback" theory, warming will turn the southern and eastern Amazon into savannah. After examining charcoal and fossil records from across the Amazon basin, researchers said that the rainforest had proven to be incredibly resilient over time to dry climatic conditions.[9] One study published in the *Proceedings of the National Academy of Sciences* argues that the Eastern Amazon will manage to attract sufficient precipitation over time and may not ultimately turn into a savannah-type environment.

The authors, however, concede that the Eastern Amazon could shift from rainforest to a seasonal forest regime given future warming. These forests could survive seasonal drought but would probably confront intensified water stress as a result of higher temperatures. Moreover, seasonal forests would be vulnerable to fires.[10] Other experts are even more pessimistic. They say that if the world's temperature rises by a mere 2 degrees Celsius from pre-industrial levels by the middle of the century then between 20 and 40 percent of rainforest trees could vanish. If on the other hand we get a 3-degree Celsius rise then up to 75 percent of the trees would disappear by 2150. How does it happen? As the planet heats up there is less rainfall in the rainforest. This in turn results in saplings failing to reach maturity while older trees simply wither away. Once the process is set in motion the cycle is difficult to stop as less tree cover results in less water evaporation, which in turn decreases rain.[11] As the rainforest gets replaced by fire-prone scrub vegetation, the Amazon would simply reach a tipping point and turn into a savannah.[12]

Meanwhile, higher carbon emissions and a globally warmer world may result in a permanent El Niño–like state that, if manifested by drought conditions, could have a huge impact on the Amazon.[13] Experts warn that the Amazon cannot withstand more than two consecutive years of drought without starting to break down.[14] If we get more frequent El Niño droughts in the rainforest, this in turn could trigger a spike in greenhouse gas emissions from burning forests.[15] By creating yet a drier climate, we may push the planet to a point where there is so much carbon in the atmosphere that the system is simply unable to extract any more.

You don't need to tell scientist Philip Fearnside about the dramatic effects of drought. One of the most cited scientists on the issue of global warming, he's been warning of the connections between extreme weather and El Niño events for years. Fearnside works with Brazil's National Institute for Research in the Amazon (known by its Portuguese acronym INPA) and sports an amusing thick moustache. A soft-spoken man, he is a long-time American expatriate and has lived in the Brazilian Amazon for thirty years. Outside his office, visitors can take in interesting exhibits about Amazonian ecology, and there are tanks full of playful manatees and giant otter.

As a young Peace Corps volunteer in India, Fearnside focused on food production and advised the Rajasthan state government on how to manage its reservoir fisheries.[16] "Have you been to Northwest India?" he asks me. When I confess that I haven't he digresses somewhat. A few hundred years ago, he says, Northwest India was lush and not desert-like. The Thar Desert, Fearnside adds, is very much a man-made desert and one can clearly see the advance of this desertification. What most concerns him is that the Amazon, too, might dry out and be placed in jeopardy as a result of climate change. In 1982, ensconced at his scientific post in the Amazon, Fearnside was shocked by the ferocity of an El Niño that resulted in forest fires that encroached on Manaus as well as the Brazilian state of Rondônia.[17]

Fifteen years later yet another El Niño caught the attention of scientists like Fearnside. From Asia to South America, El Niño spurred drought and forest fire, which spewed carbon into the air and endangered wildlife and human health. In Southeast Asia, forest fires released billions of metric tons of carbon dioxide and the region was blanketed in a choking haze. The smoke was credited with causing a spike in global temperatures.[18] Indonesia, meanwhile, was particularly hard hit—the country had not been struck by such a strong El Niño in fifty years.[19] Like Peru and Brazil, Indonesia was one of the few countries that still had vast swathes of tropical jungle—approximately 10 percent of the world's total. On the island of Borneo in East Kalimantan, 2.6 million hectares of forest was burned (and Indonesia is already the third largest emitter of greenhouse gases as a result of deforestation). In the wake of the fires, smog and smoke inhalation became a major public health concern as people complaining of respiratory problems clogged local health facilities. The drought led to food shortages in many provinces, and in the district of Irian Jaya hundreds died from drought-related disease.

Not only have we maximized the danger to ourselves through the carbon conundrum, but increased emissions now put tropical wildlife at risk. According to conservationists, climate change is altering the habitats of many primate species, leaving those with small ranges even more vulnerable to extinction. Indonesia is one of the world's three richest countries for primates, including the endangered orangutan. A mostly fruit-eating primate, the orangutan, sometimes referred to as the "red man of the jungle," plays a vital ecological role as it disperses seeds, which helps regenerate more fruit trees. In order to survive in relatively fruit-poor forests, orangutans are distributed rather sparsely throughout the rainforest. Known as the loners and daydreamers of the great apes, orangutans tend to be solitary. Indeed, orangutans may sit in the forest canopy for hours until hidden fruit seems to mysteriously reveals itself. Only then will the orangutan swing into action, grabbing its meal. Playful animals, orangutans have also been known to watch villagers use boats to cross the local waterways. Mischievously, the apes then untie a boat and actually ride it across the river on their own.

Like other fruit-eating animals, orangutans are accustomed to tracking irregular fruit supplies over vast areas. In Borneo, orangutans store fat during times of scarcity and gorge themselves when fruit is plentiful. Indonesia's unique climate plays an indispensable role in this environmental picture. Orangutans are adjusted to forests that fruit in response to Indonesian weather patterns, which in turn form part of global climate systems. Unless there is an extended drought associated with El Niño, every month of the year receives rain. During typical years, two weeks do not go by without rain. On the other hand, as El Niño becomes more frequent, extended droughts also come more often. The droughts in turn are associated with changes in the flowering and fruiting patterns of trees upon which orangutans depend.

For people living in advanced industrialized nations, protecting the orangutan and Indonesia's rainforests might seem like an abstract concept. It is anything but. Protecting orangutan habitat, especially in peat swamp forests that contain significant carbon sinks, will lead to a more secure future for the orangutan and also help to avoid carbon emissions from the forest. Conserving the Indonesian rainforest has become an ever-more-urgent priority. In Kalimantan, the Indonesian portion of the island of Borneo, orangutans suffered acutely during the El Niño drought of 1997: As the forest burned down, the creatures had little to eat or drink and faced mass starvation. Infants who were too weak to cling to their mothers fell to their deaths from trees. In desperation, orangutans fled the forest and entered local villages, where they were sometimes killed. According to conservationists, at least one thousand orangutans died directly as a result of the fires and 1000 more were negatively affected when fruit trees failed to fruit later during the wet season. Additionally, large numbers of fruit trees were simply destroyed by the fires, which left orangutan populations with a long-term food shortage.

At the same time, far across the Pacific, the El Niño drought also menaced primates in Brazil. In the Northern Amazon, forest fires ripped through the rural and inaccessible Brazilian state of Roraima. As fire penetrated savannah reservations belonging to the Macuxi and Wapixana tribes, cattle starved and produced little milk.[20] In all, 12,000 cattle were reported to have died and up to 20 percent of cultivated land was destroyed in Roraima.[21] As a result of the drought, thousands of Indians faced dire water and food shortages. Usually, when subsistence farmers carried out slash-and-burn agriculture their controlled fires wouldn't progress beyond the savannahs due to heavy air moisture. This time, however, with the El Niño drought, it was different. The fires spun out of control and soon enough the blaze entered the homeland of the Yanomami Indians. Indigenous people who lost their crops found themselves literally competing with monkeys and jaguar for dwindling food in the forest. Panicked, the animals had fled from areas where water had disappeared. As images of indigenous families driven from their homes ran on prime-time television, the fires became an international emergency. In all, more than 15,000 square miles of forest—an area twice the size of Massachusetts—caught fire during the drought.[22]

Long before we let our carbon conundrum get out of control, the Yanomami Indians developed a myth that a great conflagration might someday destroy their world. Hundreds of years later, the Yanomami believed their day of reckoning had arrived when the 1998 El Niño drought and fire hit their lands. Once the smoke cleared, the Indians thought, the sky would fall and they would all die. Though the Indians were accustomed to their own small fires, they had never seen anything so massive. In the smoke-choked village of Tototobi, five Indian elders known as *xapuris* went into a hallucinogenic trance after snorting powdered bark from the virola tree. "Cool the sun," they chanted.[23]

Few heard their pleas. Government officials, disinterested in climate problems afflicting the rainforest, declared that "the Amazon forest does not catch fire" even as blazes burned out of control for weeks. Finally, the government was forced to reckon with the situation and requested a $15 million loan from the World Bank to combat the fires.[24] Though authorities agreed to send 180 forest firefighters to Roraima, they could not decide whether to rent airborne fire-fighting equipment. Brazil, despite its huge forests, had no specialized water-carrying planes or helicopters. As the fires raged, environmentalists grew outraged. It was incredible that the world was sitting back and watching the rainforests burn, they argued. Indeed, some wondered how much worse the situation would be allowed to get before the international community acted.

Abandoned by their government and scared by the thick smoke, the Yanomami rushed to local airstrips hoping to catch planes out of the area and therefore avoid the destruction. One indigenous leader remarked, "The fires are taking over our land, killing the animals we hunt and the birds in the trees."[25] As the fires burned, airports had to be closed and hospitals became thronged with children suffering from respiratory ailments. The drought resulted in other public health problems as well. When pasture land dried up water became stagnant, resulting in greater numbers of malaria-transmitting mosquitoes. Tropical diseases like dengue reportedly flourished, and up to 80 percent of the population in some villages contracted malaria.[26]

The carbon curse and El Niño–driven drought, spurred on by wealthy nations, spread from Brazil to Guyana, where rainfall decreased dramatically and nearly all lakes, reservoirs, and other irrigation sources dried up. Drought and forest fires ruined Guyana's economy, depriving many indigenous peoples of their livelihoods and prompting the president to declare a national emergency. Guyana, already one of the poorest countries in the Western Hemisphere, could ill afford to take an economic hit. In the southwest of the country, thousands of Amerindians watched helplessly as drought and fires destroyed crops and livestock suffered. Those who relied on subsistence agriculture were hit particularly hard. A report issued by the Red Cross noted, "Food supplies, already in short supply in the hinterland, are almost exhausted and malnutrition is expected to re-emerge in many vulnerable areas, in particular among women and children." As rivers, creeks, and ponds shrunk or dried up completely, water shortages were reported throughout the country. Because of depleted water in local rivers, transportation was severely hindered. As in Roraima, Guyana was hit with outbreaks of malaria and dengue.[27]

As if El Niño–driven drought from the Pacific was not bad enough for the Amazon, the rainforest may now have to contend with other meteorological changes in the Atlantic. Indeed, researchers believe that a drought that hit the Amazon in 2005 was linked to warming sea surface temperatures in the tropical North Atlantic Ocean—the same waters that unleashed Hurricane Katrina.[28] The drought lasted from mid-July to mid-October 2005 and resulted in an environmental disaster in the tri-national region of Madre de Dios, Peru; Acre, Brazil; and Pando, Bolivia (otherwise known as the MAP region of southwestern Amazonia).

Scientists believe that the North Atlantic region will continue to warm, which in turn will suppress this July–October rainfall in western Amazonia. According to one model, increased greenhouse gases will lead to ongoing drought-like conditions. The 2005 dry spell was an approximately 1-in–20-year event, but will become a 1-in–2-year event by 2025 and a 9-in–10-year event by 2060.[29] A long-term legacy of the drought is the damaged rainforest, which has become more prone to future burning. In 2005, parched trees turned to tinder. The problem, according to scientists, is that increased tree mortality led to more dead and dry material and reduced leaf area. This in turn resulted in greater penetration of hot sunlight to the forest floor.[30]

From the orangutan of Indonesia to the Amazonian rainforest, wildlife struggled to adapt to the carbon-fueled drought. Along the Las Piedras, a tributary of the Madre de Dios River, millions of butterflies swarmed on exposed banks of riverbeds, a spectacular migration spurred by the historic drought. Most of the insects were white-, orange-, and lime-winged sulphurs but there were also tens of thousands of small brown nymphalids known as ruddy daggerwings. Though large swarms were not uncommon at the end of the dry season, people had never seen anything like this particular spectacle. At dusk, the rainforest was silent and free of the normally riotous birdsongs all around.[31]

Even without intense drought, the Amazon can be brutal for those who are not accustomed to the local climate. Iquitos is a sweltering port city located in the northern Peruvian Amazon. Its streets are clogged with tens of thousands of rowdy motor rickshaw taxis. I found the heat here oppressive; indeed for most of the day I had to recline indoors with my face close to the fan. In 2005, life in this medium-sized city of some 500,000 was disrupted. The Amazon River, which comes up to the port, dropped to a record low. Meanwhile, the city experienced long delays in the delivery of food because the river became difficult to navigate. When the water level of the nearby Nanay River went down, Iquitos had to restrict water among its citizens. People started to wonder how it was that an area with so much easy river access could run low on its water supply—a strange paradox.

If residents of Iquitos were befuddled, across the border in Brazil the people never knew what hit them. The Acre River dried up almost entirely, forcing cities that depended on the river for their water supply to do without. Indeed, the Acre River essentially became a small brook.[32] The 2005 drought, the worst in more than forty years, gave rise to wildfires and dried-up streams. People were greeted to a surreal sight when they witnessed thousands of rotting fish lying along the dry banks of local rivers; the fish were promptly eaten up by hordes of vultures. Smoke pollution, the result of a prolonged dry season and human-initiated fires, affected more than 400,000 people. At one point, smoke reduced ground visibility in Acre to less than 800 meters, prompting the regional government to create a "situation room" staffed by civil defense coordinators and scientists to deal with the crisis.[33] On some days, residents took to wearing masks when they went outdoors.

Pando, a district in Bolivia where 72 percent of the people lived in poverty, was poorly equipped to confront the environmental catastrophe. As the water level fell in many rivers, Amazon communities were cut off and found they had no means to communicate with the outside world. With fires raging, government officials declared a state of emergency and carried out firefighting efforts. However, lack of government funds complicated relief efforts. When the local water supply dipped to its lowest point in fifty years, agriculture and cattle ranching were put at risk. Authorities reported that the drought was the most severe since 1963, and even that dry spell was not as bad as 2005. As large quantities of fish and aquatic mammals perished and decomposed along riverbanks, the government grew concerned about the lack of potable water and the possible spread of disease.

After river levels fell, Brazilian peasants were marooned at sleepy market towns and were unable to return to their lands. Living with his wife and seven children in a canoe beached beside a pier, one man remarked remorsefully "We're stuck here until the lagoon fills up again, living off charity and whatever make-work I can find. We had to abandon all our crops, so I don't know what it's going to be like when we eventually go back." The man was not alone: As exposed riverbanks were turned into dunes, which were whipped into thick sandstorms, residents were left stranded. Local residents, who depended on riverways to move around, were unable to find food or sell crops. In all, 300,000 *caboclos*—river people who survive mostly on what they farm, fish, and hunt—were stranded up dry rivers and floodplains.

With local residents marooned, the Brazilian government was obliged to spring into action: Army helicopters air-dropped sacks of rice and beans, medicine, and clean drinking water to remote villages where malaria was rampant and children were dying of diarrhea. As water levels dropped, areas where the river normally flowed free instead became stagnant pools—an ideal breeding ground for mosquitoes. As a result, malaria, which had always been a problem in the region, became even more prevalent and further strained limited health-care resources. Normally, many communities relied on rivers to carry away human waste. With decreased water levels along major rivers, however, sewage began to back up and the risk of a cholera outbreak increased.

One peasant farmer spoke of his plight as an army helicopter delivered supplies to displaced residents: "With the fish all dead and our watermelon and banana crop all rotted, we'd be starving if it weren't for the food packages the government brings." With river commerce at a standstill, authorities in many areas of Brazil and Peru were obliged to declare states of emergency.[34] In the port of the Brazilian city Manaus, pontoons of floating docks sat exposed on dry land while boats lay stranded in cracked dirt. When ships carrying fuel from Manaus to power generators in the capital of the neighboring western Amazonian state of Rondônia scraped along riverbeds, the vessels were obliged to turn back. Instead, shippers had to bring supplies in by trucks via the far longer and more expensive southern land route.

What is the likely long-term impact of drought in the Amazon? Recent research on the calamitous events of 2005 provides sufficient reason for concern. In the wake of the drought scientists observed an increase in tree mortality and a reduction in growth, evidence that the forest was emitting more carbon dioxide than it was absorbing. Indeed, prior to the drought the Amazon absorbed approximately 1.7 billion tons of carbon dioxide per year. Once drought set in, however, the rainforest released the equivalent of 5 billion extra tons of carbon dioxide into the atmosphere.[35] To put it all in perspective, the total impact of the drought exceeded the annual emissions of both Europe and Japan combined. Scientists have found that the 2005 drought reversed decades of positive carbon absorption. In recent decades tropical rainforests have been a great boon to the planet, absorbing a full 20 percent of global fossil fuel emissions, a trend that was sent into a tailspin in 2005.[36]

How do certain scientists wind up studying such a forbidding and physically challenging environment as the Amazon? For some, the rainforest can be a true calling. Janet Larsen works at the Earth Policy Institute, located a mere stone's throw away from Dupont Circle in Washington, D.C. In high school, she traveled to the rainforest in Belize. "It was one of those true wakeup calls," she remarks. As her plane was descending, she saw that the rainforest had been set ablaze by farmers keen on clearing the land. Later as a young student, Larsen did a summer fellowship at the Tombopata Research Center in the Peruvian region of Madre de Dios, the same area that had been devastated by drought in 2005.[37] A Spartan yet comfortable jungle center, Tombopata was built to attract tourists and researchers alike. Because of its remote jungle location and close proximity to endangered wildlife, it's an ideal spot for scientists. To get to the center, you must first fly to Puerto Maldonado from Lima or Cusco, then continue by truck to the Infierno River Port, then board boats for a two-and-a-half-hour trip to Refugio Amazonas. From there, it's another long four hours upriver.

Hardly deterred by the arduous trek, Larsen headed out to Tombopata in order to study the poison dart frog, which in shades of yellow, gold, copper, red, green, blue, or black is one of the most colorful amphibians in the world. The frogs' elaborate designs and hues are deliberately ostentatious to deter potential predators. Larsen stayed at Tombopata for four months, living in an open-air bunk room with mosquito nets. There was no electricity, just a small solar panel that powered her research laptop. Though there was no telephone, Internet, or postal service, Larsen sometimes made radio contact with another lodge down the road. The particular frog species studied by Larsen wasn't as poisonous as other dart frogs, but these amphibians are known to be some of the most toxic animals on Earth. Indeed, dart frogs may store enough venom to kill ten grown men. On the other hand, these amphibians may hold a lot of practical utility for humans: Currently scientists are interested in exploring possible medicinal uses for some poison dart frog venom, and have already developed a synthetic version of one compound that looks promising as a painkiller.

Unfortunately, global warming may be imperiling these beautiful amphibians: Currently tropical frogs like the one studied by Larsen are crashing around the globe. "Frogs," Larsen tells me in her comfortable Washington conference room, "are regarded as one of those bellwether species that can be indicators of environmental change." Because they're amphibious, frogs need relatively good water and air quality. They breathe through their skin and are particularly sensitive to contaminants, toxins, and changes in ultraviolet light and temperature. "Because of climate changes and changes in solar radiation, plus habitat fragmentation and invasive species," Larsen adds, "frogs are more susceptible to certain forms of disease or fungus."[38]

Perhaps in an ideal world, frogs would literally be able to slither away from man-made environmental problems. Unfortunately, such is not the case: While the verdict is still out on what, exactly, is causing the demise of amphibians across the Earth, one recent study published in the journal *Nature* claims that our carbon conundrum is to blame. Global warming, the researchers argue, has already wiped out a slew of frogs—as many as 112 species have disappeared since 1980—and climate change could lead to yet more extinctions. South and Central America have fared particularly poorly: About sixty-five species of frogs have disappeared in the region. While disease is the bullet that kills the frogs, researchers say climate change pulls the trigger: Scientists have now pinpointed a link between rising tropical temperatures and the spread of the deadly chytrid fungus. The disease grows on the amphibians' skin and releases a toxin. It then attacks the epidermis and teeth, killing the frog. As global warming gets worse, increased temperatures result in more water vapor in the air. This in turn forms a cloud cover that leads to cooler days and warmer nights, which favors the growth of the fungus.

Working in the Tombopata Reserve, Larsen observed that tropical frogs were under stress. As she studied the parental care patterns of the poison dart frog in the bamboo forest, she noticed that the amphibians laid their eggs in the soil in small clutches. Watching the animals, she saw that typically one of the parents would wait around until the tadpoles hatched. The tadpoles would then crawl up on the parent's back and the parent would carry them back to a stock of moist bamboo. There, the tadpoles would mature inside water-filled bamboo stalks and later emerge as frogs. It was a drought year in Tombopata, and so Larsen was able to observe what happened as water became scarcer and scarcer. The young scientist peered into giant, three-foot-diameter bamboo stalks to evaluate tadpole counts and frog species diversity. Early on in the season, when rain was plentiful, she would always find a single tadpole or a frog in a stalk. But as the forest dried out and moist stalks disappeared, she noticed four or five frogs would take refuge in one of the remaining wet stalks. Because of this crowding, she concluded, the frogs were definitely stressed. Larsen also found that the tadpoles were cannibalistic and would eat one another. Typically the parent frog would only put one tadpole in a stalk, but if there were only one or two stalks that actually had liquid in them the parent would have to place the tadpoles together. "This shows how," Larsen tells me, "in the event of prolonged droughts or changes in the weather the frog would be a very vulnerable species."[39]

As the world comes to grips with climate change, governments and scientists will have to reckon not only with plummeting amphibian populations but also the ongoing threat to freshwater fish. According to the World Wildlife Fund, fish are increasingly menaced by the effects of climate change as temperatures rise in rivers, lakes, and oceans. According to a report issued by the organization, warmer water will translate into less food, less offspring, and even less oxygen for marine and freshwater fish. Unfortunately, hotter temperatures stand to stunt the growth of some fish, resulting in fewer offspring. Normally, fish metabolisms speed up as temperatures increase. However, insufficient food supplies could slow fish growth and reproduction rates. "To make matters worse," adds the World Wildlife Fund, "freshwater fish may not have enough oxygen to breathe as waters grow warmer. Fish filter oxygen from water, but the amount of oxygen dissolved in water decreases as temperatures rise."[40]

Once again, poor countries like Peru stand to suffer. During the 2005 drought, the local fishing yield in Iquitos increased initially, only to plummet later when water disappeared. Juan Berchota, a Jebero Indian from the Peruvian Amazon, is a forest engineer based in Iquitos. "I'd never seen anything like the 2005 drought in my life," he says. At the time of the drought Berchota found himself across the border in Brazil. There, he observed tons of rotting fish along riverbanks.[41] The Amazon is home to countless species of fish and aquatic wildlife, including the infamous piranha as well as electric eels, stingrays, and freshwater dolphin. In all, the Amazon Basin is thought to contain 5,000 species of fish that swim in its many streams, rivers, and lakes.

In Iquitos, 70 percent of the fish consumed come from a nearby reserve called Pacaya Samiria. According to Minister of Environment Antonio Brack, Pacaya Samiria contains an important fish called *paiche*. An endangered species known for its flaky meat, the *paiche* is a kind of throwback to prehistoric times and has armored scales and a flat head. Unfortunately, because of its delicious boneless meat, *paiches* have been nearly wiped out by overfishing. To make matters more difficult, the *paiche* is particularly susceptible to climate change and drought, which has caused a huge drop in the fish's reproductive cycle.[42]

Moreover, other animals that form part of the local river ecology are placed in jeopardy as a result of climate change. The tapir, a sizable animal that can grow up to 600 pounds, eats aquatic plants and grasses that disappear at times of drought. The capybara—a pig-sized amphibious rodent that lives on riverbanks—has a similar diet. The largest rodent in the world, the capybara resembles a huge guinea pig and can weigh up to 140 pounds. If capybara numbers should decrease, this could have a ripple effect on predators farther up the food chain, like the jaguar.

One of the most outlandish creatures on the planet, the shy and retiring manatee is my favorite Amazonian animal. First described as a cross between a seal and hippo, the manatee has a wonderfully round body, mostly black skin the texture of vinyl, a bright pink belly, a diamond-shaped tail, a cleft lip, and a unique sixth sense. Living life in the slow lane, manatees are fond of doing nothing much at all. However, the manatee does eat a lot of aquatic vegetation. An exclusive vegetarian that feeds on water lettuce and hyacinth, the animal eats 10 percent of its body weight in a single day. Not surprisingly manatees are robust—they can grow up to ten feet long and weigh nearly a ton.

In 2005 observers were saddened by the sight of dying manatees lying in local rivers. According to the Brazilian environmental agency, more than one hundred of the rare aquatic mammals may have died as a result of the drought.[43] In addition to climate change, the manatee faces a number of other threats, including rainforest destruction, dam building, and accidental drowning in commercial fishing nets. If that were not serious enough, the manatee is hunted for its meat and oil, which has taken a toll. It's impossible to say how many manatees are left, perhaps fewer than 10,000 in the Amazon region.

All my life I had dreamt of seeing a manatee, and in Iquitos I finally got my chance. At an installation belonging to the Institute of Peruvian Amazon Research outside of town, I saw a couple of caretakers throwing lettuce into open-air tanks. In the first tank, a pint-sized baby dolphin was scratching its back against a wood pole. Walking further, I spotted another tank full of juvenile manatees, and, in a third, a baby manatee all on its own. By the look of it, the infant was not doing too well. One of the staff explained that the caretakers found the animal in a severely mal-nourished state. While seeing the manatees was certainly one of the high-lights of my trip, it was disappointing to observe the infant in such a poor state of health.

The plight of the manatee highlights a broader problem: In an era of increasing temperatures and drought, what is to become of the world's aquatic mammals? Already, many poorer countries have seen their cetacean populations plummet. In China, for example, droughts and cli-mate change have emerged as new threats facing the Yangtze dolphin, also known as the *baiji*, and the finless porpoise. That spells bad news for these remarkable creatures, which were already menaced by commercial development and fishing. Rare cetaceans known only to the Yangtze River, the two species embody the area's biodiversity. Today, the number of finless porpoise has dropped to about 1,800, half the number found in the 1990s. In 2006, an expedition did not uncover any *baiji* in the Yangtze. For millions of years, the shy "Goddess of the Yangtze" had roamed the rivers, locating fish through use of its echolocation. Dr. Jay Barlow, a member of the *baiji* expedition, grew emotional when dis-cussing his findings. "I was stunned. I knew the species was in trouble, but I did not think they were already gone. We really had not seen the ex-tinction of a large mammal species in 50 years, so we grew complacent."[44]

Recently, as scientists have intensified their study of the effects of climate change on humans, a common theme has emerged: The Earth is an ex-tremely unequal place. In general, the people who are most at risk from global warming live in the nations that have contributed the least to the atmospheric accumulation of carbon dioxide and other greenhouse gases. The most vulnerable countries, such as Bolivia and Peru, tend to be the poorest. Battered and pummeled by drought spurred by El Niño and warming temperatures in the Atlantic the governments, peoples, and wildlife of South America are poorly equipped to withstand the ravages of climate change. The rich countries, which face the least harm from climate change, are best prepared to deal with environmental change. As one scientist put it succinctly "The original idea was that we were all in this together, and that was an easier idea to sell. But the research is not supporting that. We're not in it together."

In part, the large, industrialized countries are more resilient because of geography. They are mostly located in mid-latitude regions with so-called "Goldilocks climates": neither too hot nor too cold. Furthermore, the rich nations have abundant thick, rich soil and a more generous growing season. These disparities have spurred Third World governments as well as environmentalists to call on the more advanced countries to recognize their environmental debt to the Global South. "We have an obligation to help countries prepare for the climate changes that we are largely responsible for," says Peter H. Gleick of the Pacific Institute for Studies in Development, Environment and Security in Berkeley, California. "If you drive your car into your neighbor's living room, don't you owe your neighbor something?" Gleick asks. "On this planet, we're driving the climate car into our neighbors' living room, and they don't have insurance and we do."[45]

But while the Global North may have an easier time coping with climate change than the tropical south, we can't afford to treat the Amazon as an isolated problem anymore. The rainforest literally drives world weather systems: The tropics receive two-thirds of the world's rainfall and when it rains water shifts from liquid to vapor and back again. During the process, heat energy is stored and released. With such a vast amount of rainfall an incredible amount of heat gets released into the atmosphere—indeed the tropics acts as the planet's main source of heat redistribution. While deforestation does not seem to change the global average of precipitation, it does modify precipitation patterns and distributions.[46]

Scientists say that if we lose the Amazon this could raise temperatures and reduce rainfall in regions as far afield as the central U.S. farm belt, the Gulf of Mexico, Texas, and northern Mexico. This in turn could have a very detrimental effect on agriculture.[47] Increasingly, then, the world's climate destiny is becoming more and more interconnected. It's an important point that has been lost on the Global North, which has been reluctant to offer the necessary adaptation funds to poorer nations. In light of the climate dangers associated with rainforest deforestation, rich countries need to stop dallying. Indeed, providing billions of dollars to tropical countries is an important and necessary insurance policy that will help us to avoid even more severe planetary climate change.

Chapter Four

CATTLE AND THE

"CARBON BOMB"

A s tropical rainforests are torn down, the world pays a huge premium on climate change. To put it in concrete terms: In the next twenty-four hours deforestation will release the same amount of carbon dioxide into the atmosphere as eight million people taking a flight from London to New York. Though it may come as a surprise for some, carbon emissions from deforestation currently far outstrip environmental damage caused by planes, automobiles, and factories, though studies are ongoing on the impact factors. Indeed, the relentless slashing and burning of tropical forests is now second only to the energy sector as a source of greenhouse gases and right now Brazil is the fourth-largest emitter of greenhouse gases on the planet. Rainforest destruction is occurring on such a massive scale that plumes of smoke are clearly visible in satellite photos. Though powerful political and economic forces within Brazil are pushing deforestation, the Global North is complicit in the destruction. In fact, it is the affluent nations, acting through large financial institutions, that fund destructive tropical industries. Even worse, these same countries buy up tropical commodities that are hastening the day of our climate reckoning.

Take, for example, the nefarious cattle industry—promoted and financed by Brazilian business but also by powerful international interests. In the Amazon the cattle sector is the largest driver of rainforest destruction, accounting for 60 to 70 percent of deforestation.[1] To put it in concrete terms: Every eighteen seconds on average one hectare of Amazon rainforest is being lost to cattle ranchers.[2] As if the carbon emissions resulting from cattle deforestation were not enough, consider bovine methane emissions (or cow farts, if you want to be less delicate). While much of the debate surrounding global warming has centered upon carbon dioxide—the world's most abundant greenhouse gas—methane, which has twenty-one times the warming potential of carbon dioxide, is seldom mentioned. Methane, which is produced by fossil fuels, agriculture, and the natural processes of marsh areas, is also put out by cows. Indeed, cows are virtual gas factories that take in fodder and release methane as well as nitrous oxide, two greenhouse gases that trap heat much more efficiently than carbon dioxide. "Frat boys have nothing on bovines," notes a recent article in the *Los Angeles Times*. That's because a single cow can belch out from 25 to 130 gallons of methane a day.

When they eat, cows digest food in four separate stomachs. The first one, called the rumen, produces the methane. Scientists are now developing a special bovine antacid that would cut down on methane production. The simpler and more obvious solution to our cow and climate problem would be for more people to adopt a vegetarian diet, but since many people are unwilling to forgo meat, researchers are bending over backward to accommodate carnivores. In addition to the bovine Alka-Seltzer, scientists are also trying to develop new varieties of feed grasses that are more energy efficient and as a result produce less methane.

Unfortunately, the methane problem can't simply be solved through special cow antacid pills. That's because our dilemma is not just about belching but manure, which also emits methane as well as nitrous oxide, also known as laughing gas. Nitrous oxide is also a potent greenhouse gas. As the world ponders its global warming conundrum, Brazil is fast becoming a significant factor in the equation. Within the South American country, cattle accounts for 29 percent of total methane production and 10 percent of the world total.[3]

In Brazil, rainforest cattle has accounted for much of the country's domestic demand in recent years. Now, the cattle and climate dilemma is becoming internationalized as the South American giant moves into the global marketplace. Ever wonder where that hamburger you just ate came from? There's a chance it might contain meat from the Amazon rainforest. The world needs more food and Brazil is more than happy to supply it.

Brazil's beef exports have soared, and the country's cattle herds have grown by tens of millions in recent years. Fuelled by massive illegal ranches, Brazil has become the world's leading beef exporter, raising more cattle than all twenty-five European Union members combined. As a result of the devaluation of the Brazilian monetary currency, the real, the country's beef has become more competitive on the world market. Other advances include improvements in infrastructure such as roads, which made shipping and packing more efficient; land speculation; and land tenure laws, which have allowed colonists to simply gain title to Amazon lands by clearing forest and placing a couple of head of cattle on the land. So far Brazil has exported most of its beef to Europe, the country's meat may have qualities that some markets view as favorable. Indeed Amazonian cattle are certainly free range, grass fed, and possibly organic, depending on your definition of the term.

How can we account for the remarkable takeoff of the Brazilian cattle industry? One reason has to do with public health. Mad cow disease (also known as bovine spongiform encephalopathy or BSE) has been linked to cattle that have been fed a diet of sheep bone meal. BSE is a frightful disease in which the brains of infected animals eventually take on a sponge-like consistency. The animal later suffers motor function changes, loses the ability to walk, and then dies. Panic began to spread around the world when it was discovered that consumption of infected beef was in turn linked to a fatal, degenerative disease in humans called Creutzfeldt-Jakob disease (CJD). In the wake of the public health scare, millions of head of cattle were slaughtered in Europe, and particularly Britain, to prevent any possibility of contagion. But unlike European cattle, Brazilian cattle are grass and range fed and their diets are supplemented by feed made up exclusively of grains. This means that the Brazilian cows are considered to be free of BSE. "Fear of mad-cow disease has led to a premium on open-range, grass-fed cattle," noted one article in the *Guardian* newspaper.[4] One Swedish importer of Brazilian beef even deployed samba dancers to local supermarkets in an effort to promote the product's healthiness.[5]

While BSE was in essence a boon for Brazil, foot-and-mouth disease (FMD) was certainly an obstacle. For years, Brazilian beef was tainted in the public eye due to its association with FMD, but now the country has the problem under control. Highly contagious, FMD is caused by a virus and primarily affects cattle, swine, and other cloven-hoofed animals. For the animal, FMD is a traumatic experience: Painful blister-like lesions and erosions grow on the tongue, lips, and mouth, as well as teats and in between the hooves. The body temperature rises to a fever in order to fight off the infection. The blisters then rupture and discharge a clear or cloudy fluid, leaving raw, eroded areas surrounded by torn fragments of loose tissue. As a result of painful tongue and mouth lesions, the animal can't eat very much and feels sluggish. Most affected animals eventually recover, but the disease leaves them debilitated. They lose body weight and have trouble conceiving or carrying their calves to term. FMD used to be a pervasive problem in Brazil but recently the World Organization for Animal Health listed the Brazilian state of Pará as free of the disease with vaccine, thus clearing the way for exports to Europe, Russia, and the Middle East.

So far so good for ranchers, but the economic expansion comes at considerable environmental cost. If you go to Brazil the deforestation is all too apparent to the naked eye. When I took a night flight from São Paulo to Manaus I was taken aback to see dozens of forest fires burning off in the distance. These fires are increasing the forest's susceptibility to recurrent burning by killing trees and thereby permitting sunlight to penetrate and dry the forest interior.

Perhaps the fires were set by ranchers. On the Amazonian frontier, it's common for land grabbers to start the blazes, which can rage out of control. Experts call certain columns of flames "chimneys," which occur when the fire moves along the forest floor and reaches a standing dead tree encircled by vines. When the fire reaches the tree, it rapidly engulfs it and moves into the forest canopy.

Chimneys and the drying out of the forest have long-term implications since the Amazon plays a very important role in the production of water vapor across the region. Indeed, about half the area's rainfall comes from moisture evaporating from the forest and plant transpiration. When the forest is converted to pasture this promotes drought by decreasing water vapor flux to the atmosphere. This means that further rainfall is inhibited. Restricted rainfall in turn results in drier forests, and drier forests are more susceptible to fire. Ultimately, this can lead to an insidious "feedback loop" as forest fires inhibit yet more rainfall.

To get a sense of what is driving deforestation and climate change in Brazil, head to Pará. This Brazilian state, through which the lower Amazon River flows out into the sea, has long been known for its prized commodities. During the colonial era in the Americas, the Dutch and English made various attempts to take over the area in order to control the pepper and annatto trade. The annatto tree, which grows throughout the tropics, grows up to about thirty feet tall and has rose pink flowers. The tree's brown fruits yield a reddish or yellowish powder that is used as both a spice and dye. Annatto (also known as *achiote*), is used to color confectionery, butter, smoked fish, and cheeses like Cheshire, Leicester, Edam, and Muenster. It is also used as a coloring in cosmetics and textiles.

These days, however, Pará's prized commodity is cattle. Indeed, thanks to cheap land Pará is now the country's fifth-largest ranching state, with millions of heads of cattle. Business is booming in Pará's cattle city of Marabá, and Bertin, Brazil's second largest meat exporter, has five packing units in the surrounding rainforest.[6] The company is thrilled that European nations have lifted the FMD ban so that the company may, at long last, export its beef.[7] Over the last few years ranching has grown by half, spurred by new industrial slaughterhouses in the Amazon. There are now more than 70 million head of cattle in the Brazilian rainforest, and cows outnumber people by a ratio of more than three to one.[8] So powerful has the Brazilian cattle industry become that Bertin and other Brazilian cattle companies have become international players and have bought out other companies in the United States, Australia, and Argentina.

Once considered economically marginal, Amazonian cattle ranching now yields a lot of money. In the early 1980s the region around Marabá was surrounded by a gray desert moonscape of failed ranches. Trees had been rapidly cut and burned to pave the way for pastureland. But grass sown in the place of the deforested trees had failed to grow and looked sparser and sparser every year. Meanwhile, gaunt cows wandered for miles between meals. No one around Marabá was making any money from the cattle trade. Although land settlers had managed to perfect methods of deforestation they were clueless about how to pursue ranching profitably. But then in the 1990s the cattle population expanded as banks became a key economic player in the equation. Realizing that the land was fast appreciating in value, the bankers told ranchers: Don't worry, your land is adequate collateral for us and we will provide you with loans. Today, ranchers in Eastern Amazonia are even turning more of a profit than their counterparts in the Brazilian south.

What are the larger forces driving deforestation and climate change? In Brazil, local ranchers are frequently allied to local politicians in an insidious alliance. Indeed, in its relentless drive to promote economic growth, the government has invested in all parts of the cattle supply chain, from farm-level production to the international market. The level of financial support has been staggering: Recently President Lula made $41 billion available in credit lines for Brazilian agriculture and livestock. A full 85 percent of the credit was directed toward corporate agriculture. One chief culprit has been the Brazilian National Development Bank, which acts as the financial arm of the Ministry of Development, Industry, and Foreign Trade. In recent years, the bank has doled out billions to a handful of global players in the cattle sector, such as Bertin, Brazil's largest leather and rawhide dog chew supplier and the country's second-largest beef exporter.[9]

Without this official government support these companies would have found it much more difficult to establish themselves in the global marketplace. In exchange for its financial support, the government gets shares in the companies. In essence then, there's a joint venture promoting cattle and climate change in the Brazilian Amazon. From an economic and political standpoint, it's very difficult to roll the cattle ranchers back: In 2008, agribusiness accounted for 25 percent of Brazil's gross domestic product. Agricultural shipments meanwhile were worth a record $72 billion and accounted for 36 percent of the country's total exports.[10]

With national politicians in tow, the ranchers have an easy time of it at the local level. Indeed, they routinely steal government land by falsifying titles and bribing registrars.[11] Here's how the game works: Land grabbers, also known as *grileiros,* settle on unoccupied land and try to establish a fraudulent claim. In Portuguese the word *grileiro* means cricket. The name is linked to the notion that land crooks like to make phony ownership documents look older than they actually are by putting the papers in a drawer full of crickets. Fraud has apparently become so commonplace that Brazil's agrarian reform agency has discovered tens of thousands of bad claims in recent years. The *grileiros's* corrupt practices are bad enough as they are, but to make matters more problematic, the land grabbers also cut down trees on fraudulently owned property and turn the rainforest into pasture. Once the *grileiro* has moved on to the land and cleared the forest, he can take advantage of laborers who have fallen into debt bondage.

In Pará, the ranchers, of course, deny that anything like slavery exists. Workers are not forced to go to the farms, they argue, and laborers are almost always properly taken care of. If abuses exist, ranchers argue, it's only a few bad apples that are responsible for the problems. The ranchers, however, are whitewashing the problem. Indeed, if they were so concerned about the abuses then they would denounce the bad apples more openly. It's difficult to see how Amazonian slavery is so vastly different from earlier variants. If the workers have no possibility of leaving, if they are prevented from doing so by armed guards, if they live in miserable conditions, and if they are charged more than they earn, then this constitutes slavery.[12]

On the outskirts of Marabá, an army of impoverished men mill about outside a local supermarket looking for work. They are ragged looking and sport old T-shirts and rubber flip-flops. They carry all of their possessions under their arms in plastic bags. Some are picked up very rapidly and transported to jungle camps or isolated farms. Those who are not recruited lie about at local flophouses, known as "pioneer hotels." True hell-holes, these dens can be cramped and mosquito ridden. Human rights activists believe that some of these unsavory establishments are actually slave houses where guests are mercilessly exploited by hotel owners and gang masters known as *gatos,* or cats. The *gatos* force the unwitting guests to work until they have paid off their debts for food and housing. Today, activists say that tens of thousands of workers live in slave-like conditions and hundreds of pioneer hotels continue to operate throughout the country.

Though it may seem difficult to believe, workers caught up in debt slavery in town are probably better off than those laborers who actually get recruited. One worker who spoke to the BBC about his experiences remarked that he and others were taken several days upriver by a *gato.* Then he was dropped off and told to clear forest. The laborers only had a rough shack to live in and the food that they'd brought with them. "The *gato* said he'd be back in a couple of weeks but he never appeared," the man said ruefully. Unfortunately, there was no way the men could escape from the ranch, as they were so isolated. Six months later, the rains came and their shack was flooded. Finally, after their long traumatic ordeal the men managed to get help and fled. However, they failed to get any compensation because the landowner did not own title to the land.[13]

There's a war in Pará, and it's bloody. Because land ownership is often unclear and plagued by corruption, violence is common on the Amazonian frontier. So-called *pistoleiros,* armed gangs, force families and sometimes even entire communities off their land. Over the last decade or so most of the rural murders in Brazil have occurred in the Amazon, and Pará in particular has accounted for the majority of murders in the rainforest (one particularly bad case occurred in 1996 when nineteen landless workers were massacred). Researchers say that a significant amount of these deaths occurred in areas that had undergone deforestation.[14]

Those who strive for environmental and social justice in Pará do so at their own peril. Take the celebrated case of nun Dorothy Stang. Born into a large Catholic family, Stang grew up in Dayton, Ohio. In 1948, after graduating from high school, she joined the order of the Sisters of Notre Dame de Namur. Founded in France at the end of the eighteenth century, the order was dedicated to defending poor people, especially women and children "in the most abandoned places." Idealistic and dedicated, Stang taught elementary school in Chicago and Arizona and spent weekends and summers working with migrant workers. The young nun quickly developed a knack for earning trust. Stang easily won over a wide range of people, including lawyers, senators, and homeless people.[15]

Working among immigrants and the poor in the United States is challenging enough, but Stang later took on an even greater task when she headed to Brazil in 1966. Stang labored as a missionary at a time when progressive practices of so-called "liberation theology" were sweeping through the Catholic Church in Latin America. Discarding their habits for jeans and T-shirts, priests left their cloisters to work in shanty towns and poor rural communities among the poor and dispossessed. In the early 1980s Sister Dorothy began to work in an extremely dangerous area, the town of Anapu located some sixty miles northwest of Marabá. A dusty settlement located on the edge of the Amazon, Anapu is known for its many chainsaw shops. There, Stang began to work on behalf of landless farmers and to advocate for sustainable development projects. She also fell afoul of ranchers and loggers by criticizing their efforts to take control of land—often through phony deeds—in order to clear large tracts of rainforest.[16] To her credit, Stang also recognized the growing link between deforestation and climate change. "When I arrived we had six to eight months of rain, we now have four," she remarked to a crew of documentary film makers. "Each year we have to dig our wells a little deeper."[17] Stang, however, wasn't simply a critic but also a devoted activist. In particular, she sought to end destructive slash-and-burn agriculture by teaching farmers about more sustainable agricultural methods.

She also helped to organize local farmers to construct a factory designed to turn their produce into flour that could supplement their modest incomes. On her trips back to the United States, Stang raised money to plant trees in areas that had been hard hit by deforestation. A woman with a wonderful sense of humor, Stang rode a motorcycle, wore hand-painted T-shirts, craved chunky peanut butter, and loved making pancakes. In recognition of her local work, the Pará government named her "Woman of the Year" in 2004 and the Brazilian Order of Attorneys gave her its annual human rights prize. Needless to say, local ranchers were hardly amused by the quirky local nun or her progressive environmental politics. Indeed, while residents referred to Stang as Dora or "the angel of the Trans-Amazonian" (a local highway), loggers and others labeled her a terrorist and even accused her of providing guns to peasants.[18]

In taking on local landowners, Sister Dorothy was challenging some powerful economic forces. Spurred on by booming logging and cattle, Anapu's population swelled to 20,000 by 2005 while sawmills proliferated.[19] Along with the economic boom, lawlessness increased and Stang grew increasingly concerned about local shootouts and violent land conflicts. Even though she knew she was putting herself at risk in this Wild West–type atmosphere, Stang went to Brasilia to provide evidence to a congressional committee of inquiry into deforestation. In particular, the nun singled out logging companies that had made incursions on to public lands. If Stang hoped that her testimony would sway large landowners, however, she was mistaken. Brazenly, ranchers released cattle on peasants' crops and hired thugs to burn local settlements.[20]

Hardly content to stand on the sidelines as the situation escalated, Stang told federal police that twelve local landowners were intimidating local farmers. For the nun, the situation had become intolerable: Local farmers had received death threats from *pistoleiros* intent on securing lands earmarked for Sister Dorothy's sustainable development projects.[21] The nun, too, began to receive death threats, but despite the danger she refused to leave Anapu.[22] Stang believed that her status as a nun would offer her a level of protection, but she may have realized that the *pistoleiros* would one day come for her. Alarmed for her own personal safety, she spoke with Nilmário Miranda, President Lula da Silva's secretary for human rights. Describing the death threats that she and others had been subjected to, Stang asked for the government's protection.

Less than a week later she found herself walking on a road in Anapu with two peasants, headed to a local settlement. Farmers on the plot had apparently received title to the land from the government, but loggers were determined to get their hands on the property.[23] As the three were trudging down the red dirt road, two gunmen who had been hired by local cattle rustlers approached ominously. Stang pulled out her Bible and read to them. The two listened for a moment, took a few steps back, and shot her six times. The *pistoleiros* were paid a mere twenty dollars each for Stang's death.[24] A local cattle ranchers' leader, Francisco Alberto de Castro, described the elderly nun as an "agitator." It was she, he claimed, who fomented problems in the region and hence she had only herself to blame for her death.[25]

In the wake of the outpouring of support on behalf of Stang and international outrage at her death, the Lula government was forced to pay closer attention to the Amazon. In a flurry of activity, police arrested suspects and deployed 2,000 heavily armed troops to the region via helicopter. Finally, the authorities clamped down on slave labor, sending teams of inspectors into the Amazon accompanied by armed federal police officers. As a result, hundreds of workers were rescued in Pará. Outside Marabá, workers were astonished when they saw government agents pull up at local ranches to free them from bondage. Workers saw the farm owners' weapons and had not dared to attempt an escape. Disheveled and wearing tattered clothing, they wandered toward government vehicles and asked in a dazed tone of voice, "Who are you?"[26]

After years of criminal impunity in the Amazon, Lula's crackdown on the ranchers gave environmentalists a ray of hope. Ecologists had long believed the president had "green" politics. Reinforcing this view, Lula declared the creation of an 8.2-million-acre reserve and a national park spanning 1.1 million acres in the Pará Amazon. For years, environmental groups had campaigned for the biodiverse-rich area, known as Terra do Meio or Middle Land, which was also coveted by ranchers. But for Lula to tackle the ranchers, the president needed an effective policing authority in the Amazon. In Brazil, however, protecting the rainforest is logistically and politically difficult. IBAMA, the Brazilian Institute for the Environment and Natural Renewable Resources, is affiliated with the Ministry for the Environment. Its job is to enforce environmental law and many staff members are enormously dedicated to their profession. Some employees have even been menaced with death threats in the course of their work.[27]

IBAMA has detailed maps displaying the location of cattle ranchers' illegal air strips. The ranches are run by absentee landlords who fly in from time to time to survey their land. Determined to challenge the ranchers' power, IBAMA personnel say they are systematically getting rid of the airstrips. Despite these successes, IBAMA has been plagued with all manners of problems. The agency is grossly underfunded and some government agents don't even have money to pay for gas.[28] Even more serious, the agency has been dogged by accusations of corruption. In the weeks after Sister Dorothy's death, waves of troops made over 300 arrests of contraband logging trucks. The busts revealed that one hundred IBAMA officials were involved in illegal timber operations.[29]

If one person was up to the job of cleaning house at IBAMA, truly contesting the power of the ranchers, and getting Brazil's deforestation problem under control it was Lula's minister for the environment, Marina Silva. A former senator who came from a family of Amazon rubber tappers, Silva's story is truly remarkable. Born in 1958 in the Bagaço rubber field in the rural northwestern Amazon state of Acre, Silva grew up cutting slashes into the bark of rubber trees. The young girl had no other option but to work: She was one of eight children, and her impoverished family depended on fishing, hunting, and rubber tapping for its very survival. In the settlement where she lived, wild game was an important dietary staple, and to this day Silva's favorite meal is *farofa de paca*, a large forest rodent roasted with cassava. Even with the entire family hard at work and complementing their diet with game, Silva's family barely scraped together a living. Indeed, the young girl sometimes had to go without food.

Laboring alongside her father, a *seringueiro* or rubber tapper, Silva collected the latex as a means of helping to provide for her large family. Every day, the young girl would walk a grueling ten miles through the jungle to score the trees and later collect the latex from which natural rubber is made. As a teenager, Silva fell sick and was told that she was suffering from malaria. But later in the Acre state capital of Rio Branco she discovered that she had been misdiagnosed and that in reality she had developed a case of hepatitis. It was surely a physical and psychological blow, but while she was in the city Silva learned to read. Like Sister Dorothy, she got involved in the liberation theology movement, which had fueled political activism within the Catholic Church on behalf of the dispossessed. Her political role model became Saint Francis of Assisi, a Catholic figure famed for his spiritual relationship with animals. In 1981 she enrolled at the university and earned a degree in history. Three years later she became a teacher.

But Silva did not receive the substance of her political education in a lecture hall. As life started to improve for her she met the legendary Chico Mendes, an environmental, labor, and land reform activist. A *seringueiro* who had championed the rights of Acre's rubber tappers and defended the Amazon rainforest, Mendes befriended the young woman. In 1984, the two co-founded a local labor union. At the time, such activity was risky as Brazil's military leadership had declared labor organizing illegal. A year later, Silva joined the incipient Workers' Party, at that time just taking off under the leadership of Luiz Inácio Lula da Silva. From Mendes, Silva learned the value of true political commitment and sacrifice. The two carried out so-called *empates*: non-violent human barriers designed to resist destructive tree felling. The direct action campaigns saved thousands of acres of forest and inspired environmentalists around the world. Buoyed by their political successes, Silva and Mendes ran for federal and state deputy, respectively. Though both were defeated, they continued their environmental campaigning together. Mendes, who was born in the rubber tapper settlement of Cachoeira, worked hard to have his birthplace declared the exclusive preserve of his fellow workers. He even organized *empates* there, which did not ingratiate him amongst powerful land interests. In 1988, ranchers had him killed.

Unfortunately for the ranchers, Mendes became a worldwide martyr and symbol for the environmental cause. While the death of her mentor came as a great shock to Silva, she persevered in her political work. Representing rubber-tapper and teachers' unions, she ran for and won a local council seat in Rio Branco. From there, she continued her ascent: In 1990 she was elected to the state assembly and two years later won a seat in the Brazilian Senate with Lula's Worker's Party. At thirty-six she was the youngest senator in Brazilian history. "I was certain the issue of the defense of the Amazon . . . would not end with his [Mendes'] death," Silva remarked, "because it wasn't Chico's movement anymore. They believed that by killing Chico they would put an end to the movement. It was a great mistake. It grew bigger and bigger."[30]

Indeed, during Mendes's time, talk of defending the forest, the Indians, and the rubber tappers was perceived as being against progress. By the 1990s, however, few voiced such concerns, and even Mendes's fiercest critics agreed that Brazil needed controlled development. What's more, even Silva's home state of Acre set innovative new environmental standards by using special land zoning for protecting biodiversity and extracting wood with strict certification. "My own term as senator, everything today in Acre, has Chico's mark," Silva says.[31] Dubbed "the forest senator," Silva became a passionate advocate for sustainable development in the Amazon. True to her humble origins, she sponsored legislation that created subsidies for Amazonian rubber producers. A prominent supporter of Brazil's biodiversity resources, she backed moves to recognize Indians' intellectual property rights over native plants. An ardent critic of deforestation, Silva pressed for greater governmental oversight of the lumber industry.

Silva's meteoric rise continued over the next several years. "A bearded former metalworker and his colorful entourage . . . assume power in Brazil today, making one of the biggest shifts of government in the country's history," remarked the *Guardian* of London in January 2003. That same year, Lula, a former lathe worker and union militant, became Brazil's first working-class president. He headed up the first left-wing government since before the country's 1964–85 military dictatorship. In a move that demonstrated his commitment to the Amazon, Lula appointed Silva to run Brazil's Ministry of the Environment.

While environmentalists praised Lula's decision, some wondered whether Silva had the necessary tools to make a difference. "The ministry . . . today lacks the structure, budget and culture to do its job," remarked Roberto Smeraldi of Friends of the Earth. "To take advantage of all Marina has to offer would require a profound restructuring of the environment ministry," he said.[32] As if Silva's challenges were not great enough, the new minister for the environment received a big shock shortly after assuming her post. Touring the Amazonian state of Roraima, she witnessed nearly 700 wildfires burning out of control. The blazes, which threatened forest reserves and Indian reservations, were sparked by farmers clearing agricultural land. However, they had quickly gotten out of control due to extremely dry conditions caused by El Niño. Speaking to the media, Silva said that climate change was the worst environmental problem faced by humanity.

But in order to confront climate change, Silva would have to challenge the ranchers and other powerful interests that had stood in the way of her mentor Chico Mendes. When she was still a legislator, Silva had taken on the ranchers when they attempted to change Brazil's forestry code. Under the law, farmers were only allowed to clear 20 percent of native forest vegetation on their lands. But the ranchers' controversial measure would have permitted them to clear up to 50 percent of forest areas for crops and grazing. Moreover, a census to be carried out by states could have allowed up to 80 percent clearing on any Brazilian farm. But when Silva and other environmental groups voiced their opposition, they were able to mobilize public opinion against the ranchers' congressional caucus and the motion went down in legislative defeat.

It was the first time that the incipient environmental movement had prevailed over the ranchers, offering an important symbolic victory. The ranchers' caucus, charged environmentalists, was the human face of rampant inequality, injustice, class privilege, and impunity that had ruled Brazil for hundreds of years. "The fight over this legislation was really between the 19th century and the 21st, over the future of the Amazon. It's important that the 21st century won," noted one campaigner.[33] Later, as minister of the environment, Silva further antagonized the ranchers by sending extra federal police and environmental agents to rainforest areas plagued by illegal land clearing and deforestation. But how could the Ministry of the Environment monitor the entire vast Brazilian interior? In addition to the Amazon, cattle ranchers had spread out into a huge region called the Pantanal (Portuguese for swamp). Covering an area touching on Brazil, Bolivia, and Paraguay, roughly equal to the size of the state of Florida, the Pantanal is the planet's largest wetland.

Wetlands, which are not just swamps but also include marshes, peat bogs, river deltas, mangroves, tundra, lagoons, and river flood plains, together account for 6 percent of the world's land. They also produce a whopping 25 percent of the world's food, purify water, and recharge aquifers. Regrettably, many have regarded wetlands as an obstacle to civilization and as a result approximately 60 percent of wetlands across the globe were destroyed over the course of the twentieth century. Experts say that wetlands have an image problem with the public, which is more predisposed to saving the rainforest as opposed to the swamp. The fact is people just don't have a good impression of wetlands despite the fact that this ecosystem provides vital environmental services.

Not as environmentally chic as the Amazon, perhaps, the Pantanal is a dazzling region in its own right, consisting of an array of lakes, lagoons, rivers, forest, and forest islands. Over half the year, the Pantanal is dry and for the rest of the time turns into a shallow lake. As such, it offers a unique habitat for thousands of different species including birds, flowering plants, mammals, reptiles, butterflies, and fish. Indeed, the Pantanal contains the greatest concentration of fauna in the Americas. The Pantanal is home to the world's largest jaguars, weighing in at a full 220 pounds. Large cats are joined by giant otters, which inhabit water holes frequented by nine-foot-long caimans, also known as *jacaré*, who glide along with a sideways stare. Eco tourists and others, however, may have a limited opportunity to take in much of this wildlife: That's because some species are in danger of disappearing. The long-snouted giant anteater, for example, which claws into anthills and sticks its two-foot tongue in up to 160 times per minute to rapidly suck up stinging ants, is endangered. Also on the endangered list are the giant armadillo and maned wolf.

As per the case of the Amazon, cattle ranchers are key perpetrators of the region's ecological problems. Pantanal's sandy soils and annual floods make for mediocre conditions for agriculture, so the land is primarily used for ranching. As ranchers plow through the Pantanal, they cut trees on higher elevations and sow pastures in the lowlands. To weigh the extent of environmental damage so far makes for some sobering reading. Consider this: Pantanal deforestation in recent years has quadrupled, with 63 percent of the forest in elevated regions and 17 percent in lowland regions destroyed. Like their Amazonian counterparts, Pantanal ranchers can be stubborn: Many say they can't be expected to abandon their land and livelihood because of concerns over the environment.[34]

While the environmental destruction in the Pantanal is associated with powerful economic forces within Brazil, it's too simplistic to say that the wetland issue is a strictly South American problem. Though the Pantanal is less well known than the Amazon, it is vitally important in stabilizing climate and temperature. Just like the Amazon, wetlands hold a whole lot of carbon—one-fifth of all the terrestrial carbon on the planet to be exact, or 771 billion tons of greenhouse gases.[35] Most of this carbon is stored in organic matter in soil. One shouldn't underestimate the environmental value of wetlands in light of our current climate crisis. If you added up all of the carbon stored in the world's wetlands, you'd find that it is equivalent to about the same amount of carbon currently found in the atmosphere.[36]

Unfortunately, such carbon may be released when the soil is disturbed. If carbon from the Pantanal winds up being emitted into the atmosphere, this will create a feedback similar to that seen in the Brazilian rainforest: As more carbon is released, global warming intensifies. Like the Amazon, which has also suffered from climate change, wetlands are vulnerable to global warming. A warming of a mere 3 to 4 degrees Celsius (5.5 to 7 degrees Fahrenheit) could eliminate 85 percent of all remaining wetlands on the Earth and the Pantanal would be destroyed.[37] In light of the stark situation, it is not surprising that scientists warn of a possible "carbon bomb" if the world does not do more to save the Earth's wetlands.[38] In light of the sobering reality, it's vital for the Global North to influence the rest of the world by example. If affluent nations fail to reduce their emissions, they will imperil the Pantanal while Brazilian ranchers and others will regard the United States and other powerful countries with deep cynicism.

If the United States and other climate negotiators were to travel to the Pantanal then they would have a better idea of the environmental stakes. As the world heats up, we are sacrificing the wonders of the Brazilian wetlands. The Pantanal, for example, provides wintering grounds for many migratory birds that summer in North America. The sheer diversity of the 650 bird species inhabiting the Pantanal is impressive: While the largest bird has a wing span of nearly nine feet, the smallest weighs in at only 0.07 ounces. One local bird, the hyacinth macaw, is the largest member of the parrot family and is colored a solid, brilliant dark blue, with bright yellow patches around its mouth and beak. As the *Pantaneiros* move in with their cattle in the steamy wetlands and deforestation increases, this unusual bird is put at risk as cow pasture replaces vital bird habitat. The destruction has had a direct impact on the macaw, which survives on a high-energy diet based on native palm tree species. Far away in the Global North, the hyacinth macaw might seem a bit abstract and remote. In a sense, however, the macaw is linked to our own climate destiny.

A highly social and faithful bird that mates for life, the macaw likes to make its nest in the *manduvi*, a remarkable tree displaying large, spiny fruits. No fewer than eighteen separate bird species have been known to burrow holes inside the *manduvi* and make their homes inside the tree trunk. Later on, a variety of other mammals and honey bees make use of the holes.[39]

As it lands on the *manduvi* the macaw hollows out the soft trunk of the tree, and in the process of enlarging the *manduvi*'s natural cavities the bird creates a lining of small woodchips and sawdust for its eggs to lean on. There's little doubt that the macaw would fail to survive without this particular tree, which fulfills a vital function for diverse wildlife in the Pantanal (unfortunately, however, another animal, the cow, also has a connection to the *manduvi:* Cattle like to eat the tree's seeds before they can germinate).

Thanks to recent conservation efforts the macaw's numbers in the Pantanal have increased. Still, scientists are worried that at the current rate deforestation could result in the disappearance of all native vegetation in the Pantanal within forty-five years, which would make life impossible for the macaw. As if the situation could get no more problematic, there's also the threat posed by climate change that is warming up the wetlands. In light of the two-pronged danger, how will we save the Pantanal? If the Global North would cut its emissions, this would certainly relieve some pressure on wetlands. But as rich countries dither it is falling to intrepid reformers within the Brazilian government to save these vital ecosystems—reformers like Marina Silva.

From her perch at the Ministry of the Environment Marina Silva had real political power. Even better, Silva had the ear of Lula: From the time that the president first helped to found the Workers' Party in Brazil, environmentalists had formed an important constituency and their programs were even adopted within the party's platform. Moreover, the Lula administration saw the defense and preservation of tropical areas such as the Pantanal as one of its top priorities. Unfortunately, bureaucratic and political infighting stood to undermine Silva's valuable work. Publicly the Workers' Party was committed to environmentalism, but behind the scenes such idealism ran into conflict with Brazil's need to increase exports such as cattle. As time wore on, Silva found herself increasingly at odds with Lula and his cabinet, who she believed were more beholden to job growth and opening up foreign markets than protecting the environment. She clashed frequently with the Ministry of Agriculture regarding the proper balance between economic development and environmental protection.

Having had enough of internal politics, the former rubber tapper finally resigned from her post in May 2008. In her letter of resignation, she remarked that she was stepping down because of the "difficulty she had been having for some time in carrying out the national environment agenda." Environmentalists were crushed by the loss of their champion. "If the government had any global credibility in environmental issues," said one activist, "it was because of Minister Marina." Not surprisingly, powerful business interests were not quite as complimentary toward Silva. "I hope the next minister is not as radical as Marina. She was an obstacle to economic development in Brazil," remarked the head of the central west state of Mato Grosso agriculture federation.[40]

With Silva now gone from the political landscape, what hope might exist for combating long-term deforestation and the cattle ranchers? Eager to prove that he was no pushover, the new minister of the environment, Carlos Minc, ordered the seizure of thousands of cattle grazing on public land in the Amazon. Minc, the former state environment secretary for Rio de Janeiro and a founder of the Brazilian Green Party, reported that IBAMA agents and police seized more than 3,000 cows grazing on a ranch illegally located within a Pará nature preserve. "No more being soft," the tough-talking Minc remarked.[41] Fearing government expropriation, neighboring ranchers in Pará complied with court orders and removed their own animals from protected areas. Casting himself as a kind of environmental Elliot Ness, Minc said he would launch operation "cattle piracy," aimed at monitoring the cattle supply chain. The plan would make it harder for illegal ranches to sell their meat. Referring to the government crackdown in Pará, Peter May of Friends of the Earth Brazil remarked "This can be a good way of at least showing the government is concerned about the contribution of ranching to the problem of deforestation." "It's an important strategy," he added, "but if they do it just once and then never do it again it will be seen as a media event."[42] While Minc's efforts should be applauded, it's doubtful that the Brazilian government can halt deforestation and climate change given its incestuous economic ties to the cattle industry.

As they survey the environmental problems besetting Brazil, many in the Global North may be tempted to label the social disaster in the Amazon as essentially a Brazilian dilemma. To be sure, some people will say, richer countries bear a lot of responsibility for climate change, but now Brazil is catching up and must do its part to control its own emissions. If the Brazilian state is hopelessly compromised by powerful agricultural interests and can't get its act together to police the Amazon and control deforestation, then perhaps Brazilian social activists are going to have to step up to the plate and exert pressure from below.

There are several problems with this argument, however. To begin with, activists in the Amazon are hardly in an ideal position to combat the ranchers. With only spotty help at best from the authorities, they are vulnerable to legions of *pistoleiros* who don't hesitate to get rid of activists. Even more importantly, however, the Global North is incestuously entwined in the Brazilian cattle industry, and the peoples of Europe, China, and North America are unaware of this connection. To begin with, large financial institutions like the World Bank have provided key backing to the ranching explosion. In 2007, the International Financial Corporation, part of the World Bank, invested $90 million in Bertin despite the fact that financiers were worried that the money might help to fuel deforestation.[43]

When the bankers provided the money they apparently threw concerns about human rights violations by the wayside: The money went toward the expansion of Bertin's slaughtering capacity at its warehouse in Marabá, one of the most violent, conflict-ridden areas of the Amazon.[44] Needless to say, such economic assistance totally contradicted the World Bank's pledge to fight deforestation. Incensed by the World Bank's behavior in the Amazon, environmentalists lambasted the large financial institution for extending a helping hand to Bertin. Stung by the criticism, the bank retreated and decided to pull its funding to the cattle firm.

It's a positive step, but now the bank must guarantee that it will not fund such damaging projects in future. In addition to Bertin, the International Financial Corporation has a long, dark history of funding livestock-based agribusiness: $732 million dollars in investment over a six-year period, to be precise.[45] If we are ever to get really serious about combating climate change in the tropics then we're going to have to reform institutions like the World Bank so that they support the environment rather than foster global warming. World leaders have already pushed for such moves, arguing that the World Bank should become a "bank for the environment" with a greater emphasis on supporting environmental projects, specifically those aimed at halting climate change.

For the Global North, getting control over large financial institutions would be a good way to start addressing climate change, though rich nations need to do much more. Consumers for one are going to have to wise up to the cattle-climate connection. That's because major international companies are driving Amazonian deforestation through their purchases of leather, beef, and other products supplied from the Brazilian cattle industry. The companies include big brands such as Unilever, Johnson & Johnson, Toyota, Honda, Gucci, Louis Vuitton, Prada, IKEA, Kraft, and Wal-Mart.[46] In addition to private companies, the public sector has also been caught up in Amazonian cattle. Indeed, other purchasers include the British National Health Service and a supplier in the Middle East linked to the British, Dutch, Italian, Spanish, and U.S. military forces.[47]

It all starts in places like Marabá. Outside the town lies a ranch that supplies animals to Bertin. Under the law, Amazonian farms must retain 80 percent of the original forest within their borders. But outside the ranch there's pasture as far as the eye can see. GPS data and satellite imagery show that only 20 to 30 percent of the farm is covered in forest. Throughout Mato Grosso and Pará there are dozens of other ranches that break the rules and supply animals to Bertin and other slaughter companies. After the animals are slaughtered, Bertin ships the meat, hides, and other products to an export facility in the Brazilian south and then onward to foreign markets.

In the Global North, consumers are contributing to climate change when they buy products derived from Amazonian cattle. Meat, for example, is canned, packaged, and processed into convenience foods. Hides, meanwhile, are used to make leather shoes. Nothing goes to waste: The fat stripped out from the carcasses makes its way into toothpaste, facial creams, and soap. Even the gelatin squeezed from Amazonian cattle bones, intestines, and ligaments is used to thicken yogurt and make chewy sweets.

In the high-stakes game of deforestation and climate change no one is blameless. Take, for example, China, which gets its leather from Bertin to manufacture shoes.[48] Chinese tanneries that receive Bertin's leather in turn supply several manufacturers working for well-known shoe brands such as Nike, Adidas/Reebok, and Clarks.[49] Italy, the center of high-quality leather production for the fashion market, also receives leather supplies from the Brazilian cattle giant.[50] What's more, Bertin supplies leather to U.S.-based Eagle Ottawa, which in turn sells the leather to auto manufacturers. If you drive a Chevy Malibu, Cadillac CTS, Honda Accord, Toyota Tundra sport utility vehicle, Lexus IS-F, as well as some Audi and Mercedes models, then you have Amazonian leather in your upholstery. Many Europeans and Americans not only have Bertin leather in their vehicles but also in their homes. That's because IKEA and Ethan Allen get their furniture from a Chinese supplier linked to Bertin.[51]

In Britain, the convenience food market and food service industry are changing consumer tastes. So-called frozen and chilled "ready meals" incorporate imported Brazilian beef provided by Bertin.[52] In the United States, meanwhile, people can't get enough of Burger King, which is also supplied by the Brazilian cattle giant.[53] It's not only human beings who are gobbling up Brazilian meat. Indeed, Bertin also exports millions of dollars worth of dog chews, with the United States comprising the principal market.[54] Bertin's ties to the fast food industry highlight a larger problem: The Global North's insatiable addiction to a meat-based diet. In light of our climate difficulties, we're going to have to reconsider our dietary choices. The United Nations Food and Agriculture Organization finds that meat production gives rise to more greenhouse gases (14 to 22 percent of global emissions) than either transportation or industry. Furthermore, beef is the most carbon-intensive form of meat production. Consider: A one-pound patty results in about 36 pounds of carbon dioxide emissions, or thirteen times the emissions from chicken.[55]

Given these disadvantages, perhaps one of the best things anyone can do to tackle climate change is to have one meat-free day a week and gradually decrease meat intake thereafter. It's not enough, however, to simply transition toward a vegetarian diet, which includes lots of milk, butter, and cheese—this probably won't reduce emissions significantly as dairy cows would still release methane through flatulence. While it may sound a bit naive to think that people will change their eating habits any time soon, such a move is certainly much less complicated than getting people to switch their mode of transport.

Changing our diet may be one way of saving the tropical rainforest and arresting climate change, but there are still other approaches that need to be pursued. For starters, international companies need to stop buying from irresponsible cattle suppliers. Fortunately, there's been some progress in this area—when Greenpeace released a report linking Amazonian deforestation with some of the world's most well-known industries, some companies got embarrassed and agreed to change their purchasing policies. Such was the case of Tesco, one of Europe's largest retailers, which sold beef and leather products with ties to Amazonian deforestation. In response to mounting criticism, the company said it would pressure its suppliers to change their environmental practices. Meanwhile Nike has released new leather sourcing guidelines designed to halt the pace of Amazonian deforestation. The problem, however, is that there's currently no enforceable certification system for industries involved in the Brazilian meat and leather supply chain.

Between Bertin and the Brazilian government there's plenty of blame to go around for the burgeoning climate and cattle mess. Yet there's also a significant international component to the deforestation dilemma. From the World Bank to destructive consumer tastes, rich nations bear significant responsibility for environmental stresses in the Amazon. As long as affluent countries fail to restructure their economic and financial relationships with Brazil, many in the tropics will look upon the Global North through cynical eyes.

Chapter Five

SOY IS KING

Think about soy and the first thing that may come to mind is tofu, a common staple in many Asian cuisines. Yet soy can also be used to enable the Global North's voracious demand for meat and that is where the world gets into even greater climate troubles. On Brazil's agricultural frontier it is ranching and soy that are driving deforestation—they go hand in hand. Soy has been hailed as a "miracle bean" because of its high protein content—40 percent by weight. It has long been popular among vegetarians but it is now prized as a quick, cheap, and safe animal feed for poultry, pigs, and cattle. Affluent countries that have already done business with the likes of cattle companies such as Bertin are now exacerbating climate change even further by buying up Brazilian soy and financing its commercial development.

Almost no one in notoriously carnivorous Brazil eats tofu—most of the national soy crop goes toward animal feed. But in the wake of the mad cow disease scare, rich countries increasingly demanded meat from cows that thrive on a soy meal diet as opposed to animal-based feed, thus benefiting Brazilian farmers. Brazil is also exporting its soy. To date, the Chinese and Europeans have become voracious consumers of Brazilian soy, catapulting the South American nation to agribusiness giant status. In China, where soybeans were first widely cultivated, soy imports went up by an astronomical factor of ten between 1999 and 2003, in large part because of growing affluence and a shift in the Chinese diet. For many Chinese, consuming meat and dairy products symbolizes wealth, status, modernity, and escape from rough rural life. Though the average American eats more than 250 pounds of meat ever year, the Chinese are now catching up and currently consume 115 pounds. Per capita consumption of pork in China has meanwhile almost doubled. Though China produces a lot of soy on its own, it is now the world's largest importer of soy to feed its growing livestock sector. In Europe, demand for soy has skyrocketed.[1]

Though the soy planters cut down some forests, their influence is often more indirect. Once ranchers have cleared land in the Amazon the soy planters buy up property and move in. But as they take up cleared land, savannah, and transitional forests, the soy magnates push others such as slash-and-burn farmers even further into the forest. Soy then acts as a significant push factor and catalyst of climate change. The farmers who get pushed into the rainforest by agribusiness quickly find that Amazonian soils are notoriously low in fertility. After several harvests, crop yields start to disappoint and eventually farmers abandon the land altogether or convert it to cattle pasture. It's all a social and environmental time bomb since there are hundreds of thousands of subsistence farmers trying to eke out a precarious existence in the Brazilian Amazon. In addition to pushing ranchers and slash-and-burn farmers into the forest, soy magnates exert pressure on the Amazon in other ways. For example, they lobby for highways and infrastructure projects, which paves the way for yet more deforestation.

There's no great mystery surrounding the precise role of these soy planters: Over the past few years environmental groups have published scores of papers about the issue and the international media has turned up the heat. And yet, putting a brake on the soy lobby has proven to be very challenging due to the latter's vast political and economic muscle. In Brazil important and well-connected people control the soy industry. Erai Maggi is a soy magnate from Mato Grosso who owns half a million acres of land.[2] A migrant from the Brazilian south, Maggi capitalized on groundbreaking technological developments. Though Mato Grosso's soils were considered overly acidic, scientists found that heavy applications of dolomitic lime and phosphate-rich fertilizers could render the land incredibly productive. For soy agriculture to flourish in Mato Grosso it was also necessary to develop a "tropical soybean." Gradually Maggi bought out his neighbors. Today the soy tycoon keeps tabs on the outer reaches of his empire by flying in a two-prop plane. Bom Futuro, his company, employs 300 combine harvesters and 500 tractors. Every year Bom Futuro produces more than 600,000 tons of soy, most of which goes to feed livestock, which in turn winds up as meat in China and Europe.

Currently Mato Grosso is not only Brazil's largest cattle ranching state but also its greatest soy producer. Erai's cousin Blairo is the head of the world's largest soy producer and is known as "O rei da soja"—the soya king. After serving as a senator for Mato Grosso, he successfully ran for governor in 2002. As governor, his key goal was to triple agricultural production in the state within ten years and to develop agro-industry. In a mad rush to plant soy, planters cleared vast areas of land, and during Maggi's first year in office Mato Grosso's deforestation increased by about 30 percent.[3] The state's economic development has come at a stiff price: Along with large agribusiness giants such as Archer Daniels Midland and Cargill, Maggi has driven the soy invasion in the north of Mato Grosso. As he buys up savannah land for soy cultivation, Blairo pushes cattle ranchers into the rainforest, where they proceed to slash and burn, in the process releasing millions of tons of carbon into the atmosphere. At least indirectly then, Maggi has accelerated Amazon deforestation.[4]

It's a bitter irony that in the long run Blairo Maggi might actually cause himself some business harm. Here's how it works: Soy magnates contribute to deforestation, which in turn exacerbates global warming. Climate change in turn gives rise to drought, which makes it more difficult for soy to thrive. In order to help farmers withstand the ravages of climate change, a number of large biotech companies, including Monsanto, are seeking patents on genetically modified (GM) crops designed to withstand drought and other environmental stresses. Monsanto is getting into the business of GM crops at an extremely opportune moment. In Brazil, farmers are worried that hotter temperatures will push crops past the country's borders, uphill into the Andes and toward the tip of South

America. Brazil's soy crop, which is the largest outside the United States, could lose approximately 20 percent of its cultivatable land by 2020. As a result, soy exports could slump by more than a quarter over the next twelve years.

In Brazil, however, GM soy has spurred a contentious political debate. To Marina Silva at the Ministry of the Environment as well as groups like Greenpeace, GM soy and biotechnology are merely the Trojan horse of multinationals. Environmentalists oppose the spread of GM soy, claiming that it negatively impacts small farmers who tend to plant their own soybeans next to GM crops that are vulnerable to Roundup herbicide. The herbicide has been used in conjunction with the advance of GM soy in the Amazon. Widespread use of Roundup has serious impacts on rainforest ecology: it kills all plants indiscriminately and leaves only GM crops intact. Monsanto is the manufacturer of the herbicide, which is used on Roundup Ready soybeans. Who produces these soybeans, which are less sensitive to glyphosate, the active ingredient in Roundup herbicide? That would be Monsanto, of course.

Peasant groups are outraged, charging that this is a corporate set up. Seeds form part of humanity's heritage, they say, and ought to be readily available to farmers without being subject to market pressures. In recent years, many poor countries that stand to be affected by climate change have rejected biotech crops as a result of environmental and economic concerns. In the long term biotech companies may be able to leverage climate change as a means of penetrating these resistant markets. Ironically enough, however, even though GM crops have been designed to withstand climate change they may wind up exacerbating it. The problem is that GM crops contribute substantially to increased pesticide use and give rise to herbicide-resistant weeds. To combat the weeds, biotech firms develop new GM crops that promote even more pesticide use. This spells danger for climate since the mechanical tillage involving hoeing, grubbing, and digging which is used to control resistant weeds is also increased and thus contributes to greater soil erosion and greenhouse gas emissions.

Concerned about environmental blow back, Silva remarked "There are people who do not want to grow transgenic soy, so they must have their rights ensured . . . This does not apply just to Brazil but it should to the whole world."[5] Silva however was on the losing side of the debate: Arrayed against her were Lula's economic team and Agriculture Minister Roberto Rodrigues. A soy farmer from the state of Maranhão, Rodrigues reportedly refused to recognize any connection between deforestation and the advance of agribusiness in Brazil.[6] Ultimately, conservative elements in the government were able to prevail and pushed through a measure lifting a ban on GM seeds, and Brazilian scientists are now busy at work developing genetically modified soy beans that can withstand higher temperatures.

If you believe all of the ecological problems stemming from the soy industry originate within Brazil, think again. Indeed, private Japanese and European banks have helped to finance Maggi's operations. Then there's the International Finance Corporation, the private-lending arm of the World Bank, already well known in Brazil for its sponsorship of the cattle industry and Bertin. In 2004, the corporation gave a loan of $30 million to Blairo Maggi for a soy expansion project in the Amazon.[7] In an interview with the magazine *Soy Digest*, Blairo said he was unconcerned about the prospect of deforestation linked to his project. "I do not feel responsible in the least," he declared. Civil society groups in Brazil were outraged by the conduct of the International Financial Corporation and protested the loan, saying it was socially and environmentally unsustainable.[8] In response to the complaints, the president of the World Bank ordered an audit of the Maggi loan.

As international financial institutions dither, Blairo proceeds full speed ahead with his soy expansion in Mato Grosso, with ominous implications for climate change. The words "Mato Grosso" translate as "dense forest," but recently this Brazilian state has been subject to some of the most rampant deforestation on the planet. Greenpeace awarded Blairo its first "Golden Chainsaw" prize—to be awarded to Brazilians contributing most to the Amazon's destruction.[9] When the *New York Times* questioned Maggi about the deforestation, the governor responded: "I don't feel the slightest guilt over what we are doing here . . . it's no secret that I want to build roads and expand agricultural production."[10]

And build he did: With support from the Brazilian and Mato Grosso state governments, as well as from private companies including his own, Maggi constructed roads and ports and expanded waterways through the Amazon. Critics say these projects have further opened up the region to soy cultivation, cattle ranches, and small farmers. Of particular concern is the so-called "soy highway," BR–163, which is currently being paved from Mato Grosso's capital of Cuiabá near the Bolivian border to the deep-water Amazon River city of Santarém in the state of Pará. The highway's supporters claim it will bring jobs and development, and they even compare the project to the historic transcontinental railroad that opened up the American West. Soy planters are excited about the prospect of the road being completed as their land and sea transportation costs stand to decrease dramatically. Traditionally, Mato Grosso farmers have had to send their soy crops in truck convoys that wind up traversing 1,200 miles before arriving at ports in southern Brazil.

The Lula government, which has been hopelessly compromised through its ties to the cattle industry, says that things will be different with the soy industry. Though Lula has stated that one of the major projects of his second presidential term will be the paving of BR–163, the former labor leader declares that Brazil can develop the region around the road without increasing environmental destruction. His proposal, the "Sustainable BR–163 Plan," involves twenty ministries and represents the country's most ambitious attempt yet to balance growth and conservation. Under the plan, the government has pledged to protect a full sixteen million acres of rainforest on the western flank of BR–163. If companies are seen to be environmentally responsible they will be allowed to set up limited logging concessions, though no clear cutting or settlements will be permitted. In keeping with the new plan, the government has declared that it will also strengthen environmental enforcement in the region. Critics, however, are skeptical. They point out that historically, roads have frequently led to environmental degradation in the Amazon. Furthermore, they point out, BR–163 is no minor deal: The highway's area of influence is a quarter of the Brazilian Amazon and will cut a 10-million-hectare swath of land through the area.

BR–163, the road that threatens to exacerbate our global climate conundrum, stands to put an end to Mato Grosso's romantic image. Lying in the geographic heart of South America, the state is geographically remote. In 1914, after he served as U.S. president, Teddy Roosevelt passed through the area, looking for an uncharted Amazon tributary. Ten years later, the English explorer Colonel Percy Harrison Fawcett mysteriously vanished in Mato Grosso while looking for the fabled lost city "Z." Trading in his pen for a pistol, in 1932 literary editor Peter Fleming, brother of the famed James Bond author Ian Fleming, traveled to Brazil to carry out a celebrated search for the disappeared Colonel Fawcett. After traversing some 3,000 miles of wilderness, including alligator-infested rivers, Fleming wrote a travel book called *Brazilian Adventure* based on his experiences.

Fleming, who failed to find any trace of the enigmatic Fawcett, remarked that "less, probably, is known about the interior of Mato Grosso than any other inhabited area of equal size in the world." He added, "There are huge tracts of jungle which no white man has attempted to enter. You can believe what you like about those regions: no one has authority to contradict you. You can postulate the existence in them of prehistoric monsters, of White Indians, of ruined cities, of enormous lakes."[11] As recently as the 1960s no decent roads or railroads connected Mato Grosso to the rest of the country. The crumbling state capital of Cuiabá was then little more than a hardscrabble town frequented by local cowboys and alligator poachers. Even today, Mato Grosso has a remote feel to many Brazilians, most of whom reside in the crowded coastal southeast. Fewer than three million people live in Mato Grosso, which is larger than Texas, compared to the more than twenty million who reside in the city of São Paulo.

Despite its geographic isolation, Mato Grosso now stands to lose its exotic mystique. The state is becoming integrated into the rest of Brazil, much as the United States absorbed the Wild West during the nineteenth century. Today Cuiabá is a city of some half a million and the capital of so-called "soylandia." New towns built on soy wealth have been caught up in the business fever and clocks in local fancy hotels even show the local time in Chicago, the center of the world's commodities market. In addition, frontier settlements are springing up in the middle of nowhere amidst soy silos emblazoned with Maggi's name. Each new town has a bus stop accompanied by a bar. Inside, male patrons consume *cachaça*—hearty sugar-cane rum—while resting their hands on women known as *meninas de programa,* or prostitutes.

Before heading north toward the Amazon, many stock up on boots and saddles in Cuiabá. During the dry season, though, passing over BR–163 can be an exercise in dodging huge potholes. Conversely, in the rainy season, when the road turns to mud, navigating the highway really becomes an adventure sport. At these times the highway becomes as slippery as glass and vehicles can roll down steep banks and into the river. Passing trucks have difficulty getting up steep inclines or lean at crazy angles. Collapsed bridges are a danger. Motorists who get stuck have to wait for a tractor to pass by and pull them out of the mud. Unfortunately, help may not arrive for hours or even days. Crowding around camp stoves, truckers share warm thermoses of coffee and heaping plates of rice and beans. Not surprisingly, there's a lot of good money to be made by hawkers who provide everything from exotic jungle hardwood to Coca-Cola along BR–163.

Over the past twenty years or so, soy plantations here have sprung up in the vicinity of the highway and now dominate the landscape. At harvest time, fleets of green and yellow combines crisscross local fields lining the highway. John Deere dealerships are proliferating. Lucas do Rio Verde, a town along the highway, is bustling with farming activity. Currently Lucas do Rio Verde County is playing an economically important role in Brazil: The area produces 1 percent of the country's soy crop, 10 percent of its corn, and 4 percent of its cotton. A far cry from coarse and rowdy boomtowns, the city even has flowered plazas. In the next few years, Lucas do Rio Verde's population is expected to double to 50,000. Within Brazilian development circles, the town is perceived as soylandia's jewel in the crown. A large agricultural supply firm called Fiagril has played a big role in Lucas's most recent accomplishment: a soy megacomplex bustling with trucks, cranes, and red dust outside of town.

In addition Blairo plans to build a processing plant that will truck in 3,000 tons of soybeans a day. Much of the soy will be processed into meal rations for chickens and swine, thus making Lucas into the next Des Moines or Decatur, Illinois. With all its high-tech know-how, Lucas has a rather confident, twenty-first century feel. A rising population of young, rural entrepreneurs, the offspring of earlier land colonists, is intent on making its fortune here. The men wear Stetsons, speed along in Toyotas, and draw their cell phones gunslinger style. The economic boom has come at an environmental cost, however. In 2006, thousands of local residents in Lucas got sick when a small, single-engine plane sprayed herbicide over soybean fields.

Further along the soy highway is Sinop. A generation ago the area around the town was rainforest. Today the place is abuzz with business activity. Sawmills, a slaughterhouse, and a soy plantation are all part of the local landscape. Alongside all this industry, however, shantytowns have sprung up, housing the land pioneers who have been left out of the economic bubble. To complement their diet, these poor residents head to the nearby Teles-Pides River, a branch of the Amazon, to fish. People complain, however, that the fishing has gone bad as a result of toxins from soy crop dusters. The Landless Rural Workers' Movement, known by its Portuguese acronym MST, has been organizing in Sinop, and members have illegally occupied local land. But the MST has also successfully moved the poor from improvised settlements in Sinop on to permanent farms. Leaving their shacks at the crack of dawn, activists show up on private land waving hoes and machetes. Then they start to build shelters with as much wood and plastic sheeting as they can carry. Struggle though they might, policemen and landowners have killed more than 1000 activists in Brazil since the mid-1980s.

Traveling along BR–163 after Sinop, one arrives in Guarantã do Norte, a town of about 30,000 people. Located near the Amazon, Guarantã has been wracked by smoky haze accumulating from countless rainforest wildfires. The pollution is creating rising rates of asthma among the children. In the local hospital doctors dole out oxygen masks and report that a whopping three-quarters of the patients are suffering from respiratory ailments. Along BR–163, honky-tonk frontier towns filled with stores, bars, and billiard saloons have sprouted up. One town, New Progress (or Novo Progresso in Portuguese), lying past Guarantã on the unpaved section of the road, is booming. From 2003 to 2008, the population doubled from 20,000 to 40,000. Ten years ago, the cattle herd here numbered 50,000, but that has now jumped to a million.[12] In Novo Progresso, businessmen sell chainsaws and veterinary supplies in compact concrete buildings stretching along BR–163. The dirt stretch of the highway itself serves as Novo Progresso's main street.

Even though BR–163 is barely passable for a good chunk of the year, tens of thousands of settlers have spread out in the vicinity of the road. At local hotels in the district of Itaituba along the highway, ruddy-faced businessmen from the Brazilian south—the descendants of German and Italian immigrants—cut land deals. But while newcomers have found that land is plentiful in Mato Grosso, virtually none of it is theirs to settle. Wealthy farmers, with the assistance of faked title deeds, corrupt local officials, and private armies have claimed vast areas. Though much of the land is federal property, farmers have cleared it and carpeted the ground with soy.

The influx of newcomers has encouraged a kind of Wild West frontier lawlessness. The federal government, meanwhile, has done nothing to halt powerful interests, and those who denounce land grabbers for evicting poor farmers sometimes wind up dead. Unfortunately, the actual physical work around the highway isn't much safer. In local towns, small cemeteries display unmarked graves—men killed in logging accidents. Many of these anonymous workers come from poor states and are known only by a nickname. Their bodies go unclaimed.

Could there be a viable alternative to the environmental dystopia of BR–163? Not everyone living in the area of BR–163 is intent on pushing soy or unsustainable corporate agribusiness. To get an idea of how the other half lives, head to the Amazonian town of Santarém. Near to the city, one side of BR–163 is bordered by the Tapajós National Forest, which spans 545,000 hectares and is home to many bird and tree species. Within the reserve, the Maguari people have demonstrated a great commitment to sustainable environmental practices that do not enhance our climate problem. Living off subsistence agriculture, the Maguari also create jewelry from seeds, bags and shoes from rubber trees, and honey from native bees. The people sell their products in Santarém and São Paulo and even export to France. Recently, the Maguari started offering eco-tours within the Tapajós and Prince Charles of England was one of the featured visitors. To its credit, the World Bank has funded forest management projects within the Tapajós forest and promoted sustainable pursuits such as cultivation and use of natural vines, production of handicrafts, and use of fallen timber for small-scale wood processing.

If we want to get our climate change problem under control, international financial institutions should halt their funding of destructive agribusiness and invest their money purely in alternative economic development schemes, such as the cultivation of Brazil nuts. From the perspective of global climate change, harvesting Brazil nuts, known to Brazilians as *castanha do para*, would be preferable to growing soy. The large Brazil nut is technically not a nut but the seed of a tree discovered by eighteenth-century European explorer Alexander von Humboldt. Unique to the Amazon basin, Brazil nuts grow in four- to six-pound pods approximately the size of one's hand, and one pod can contain up to thirty seeds. The tree itself is truly remarkable: Standing at 165 feet, it only starts to bear fruit at thirty years old and may live between 500 and 800 years. For centuries Brazil nut trees have grown wild in the Amazon and indigenous tribes such as the Yanomami used the nuts to supplement their diets. Rich in vital nutrients, Brazil nuts contain omega–3 fatty acids as well as selenium, calcium, and magnesium.

Collectors, or *castaneros,* must wait until Brazil nut pods drop to the ground to gather them. Later, the harvesters carry the nuts by river or road to processors. Currently there's a huge market for Brazil nuts in Europe and North America, and tens of thousands of tons of the shelled and unshelled nuts are exported abroad every year. Unlike soy, which spurs deforestation and climate change, harvesting Brazil nuts is economically and environmentally sustainable. Indeed, a tree which is over 400 years old, may ultimately provide prosperity for generation after generation, and Brazil nuts are considered one of the most valuable products that can be harvested from the undisturbed rainforest. In Brazil the nuts constitute a sort of currency and are even traded for food and manufactured goods.

It all sounds idyllic enough, though unfortunately soy interests threaten to undermine this picture. Within the vicinity of BR–163 the land has been laid waste by large soy plantations, and pastures and farmers who harvest nuts from the forest complain that the unrelenting clearing has depleted their food supply and livelihoods. Brazil nut trees are very sensitive to fires set by land grabbers, and once they're gone there's little chance the trees will return, since they do not have an easy natural regeneration and saplings don't grow in the shade. Hopefully the nut collectors and sustainable economic and environmental development will prevail, though recent developments have not been promising. Confronted by the recent land grab in the vicinity of BR–163, many poor farmers wind up selling out to developers. When they're offered 100,000 reais (about $50,000) residents relinquish their property, as they think it's a lot of money.[13] Farmers in the town of Belterra, located one hour's drive south of Santarém, traditionally made a decent living on their small plots by growing such staples as corn, squash, and beans. In the late 1990s, however, Brazilian southerners began to buy up local land at a pittance. Figuring that their economic troubles had come to an end, farmers left for the city. Later they realized they were short on funds and wound up going hungry. In Santarém, former farmers from Belterra join others in the city's growing slums.

Those who refused to sell their land did not fare much better. Encircled by encroaching wasteland, raging chainsaws, and huge blazes, the farmers watched helplessly as their land was consumed by vipers, bees, and rodents fleeing the destruction. Near BR–163, loggers illegally cut down trees used by local Indians to construct their canoes.[14] As tractors sprayed cleared fields, toxic clouds of pesticides filtered into nearby settlements. It wasn't long before people grew sick and their pigs and chickens started to die.[15] Peasants who refused to sell out to land speculators were visited by *pistoleiros* who set fire to houses and sadistically forced local residents, including children, to watch.[16]

In Brazil, those who question the role of unsustainable development in climate change—people like Ivette Bastos—pay a high personal price. A union organizer, Bastos already knows something about social conflict in the area of BR–163. Born and raised in the small village of Carariaca, two hours by boat from Santarém, Bastos grew up among subsistence *caboclo* farmers (in Brazil, *caboclos* or *mestiços* are people of mixed European and Indian ancestry). In villages like Cariaraca, there's a lot of poverty but people don't starve as the rivers provide fish and the trees are full of fruit. Growing up with her parents and eleven brothers and sisters, Bastos picked fruit to sell in local markets. There are more than a few parallels between Bastos's life trajectory and that of Marina Silva. Like Brazil's former minister of the environment, Bastos worked as a rubber tapper in order to raise cash. Later at school she became a skilled community organizer and got involved in a local church group. When she hit her thirties Bastos became more politically active and joined the Santarém rural workers association. As Cariaraca's local representative in the association, Bastos pushed for the installation of a public telephone in her village. Efficient and popular, she was made regional director and moved to Santarém.

Popularly called the "Pearl of the Tapajós River," Santarém is the second most important city in the state of Pará. The town is surrounded by forests; *igapós,* or flooded wood areas; and many ponds and *igarapés,* or small river lagoons. One of Santarém's main attractions is the meeting of the Tapajós and Amazon rivers, whose greenish-blue and muddy waters, respectively, won't mix together. Though Santarém is a sleepy city, its laid back atmosphere could be turned upside down by the soy boom. Indeed, it's probably only a matter of time before the remaining road between Guarantã and Santarém is finally paved, and when the government announced plans to complete this stretch of BR–163, a veritable land grab ensued. For people like Bastos, soy has brought only misery, fear, and exploitation to Santarém. In 1999 the town became a social and political flashpoint when agribusiness giant Cargill decided to construct a grain terminal here. When the facility opened at long last in 2003, the region was transformed.

Outside Santarém, subsistence farmers and *caboclos* had little experience with mechanized agriculture. But now, excited by the prospect of massive exports, developers began to move in and plant soy on a large scale. A full 300 square miles quickly became covered in the crop. In the long term this soy production could prove socially and environmentally unsustainable. Soy is a monocrop, needs large areas to be economically viable, and destroys biodiversity. Moreover, soy encourages greater concentration of land ownership in the hands of a few. Finally, people cannot find work on soy plantations as local farmers are uneducated and the only jobs available on soy farms are skilled machinery operators.

For Bastos, moving to Santarém came as a big cultural shock. One night she left her flip-flop sandals outside her house as she might have back in the jungle. The next morning, she found they had been stolen. Aside from such problems, Bastos faced great challenges in Santarém at her new job. Her arrival in the city coincided with the onset of the soy boom and many communities found themselves under threat from land speculators. By now elected president of her rural association, Bastos sought to dissuade people from selling their land. Holding meetings and organizing demonstrations, she was now the most prominent local activist campaigning against soy in the wider region. For Bastos the heart of the problem was the so-called *grande,* or political and economic, elite: *grileiros,* soy producers, ranchers, and loggers. Determined to halt the soy expansion, Bastos and her colleagues sought to get the authorities to provide proper titles to those who already resided on the land.

Others, however, wondered whether it might not be more effective to lobby the government to simply dole out bullet-proof vests. Predictably enough Bastos started to receive death threats as a result of her work. On a visit to a local health clinic, a doctor remarked that he'd seen her on television. "I hear you're going to die," he declared. On her trips to the rainforest she was frequently stopped by local men who would deliver sinister messages. The men told her that if she were ever to return to the area it would be in a coffin. What's more, if she ever managed to achieve land rights for local subsistence farmers she would be turned into ashes. Bastos started to travel with bodyguards and would not set foot on large soy plantations for fear of retaliation. One of her colleagues was kidnapped, thrown in the back of a car, and then taken to the forest. Once there, he was tied to a tree for the entire day. The pressure took a physical and emotional toll on Bastos: She began to suffer from stress and went through serious bouts of depression. During such stretches, Bastos headed to Carariaca, the only place she felt safe, for some rest and relaxation.[17]

It's hard for Bastos and others to challenge the soy juggernaut when the industry gets multi-million-dollar funding from international financial institutions, and business magnates have high-up political connections. But the Global North is mixed up with Amazonian deforestation and climate change in yet other nefarious ways. Popcorn chicken, according to the KFC website, is "bite-sized pieces of all-white meat chicken, marinated for a tender inside and breaded for a flavorful, fun crunch on the outside." A great snack for adults, popcorn chicken is also "a fun way to please picky kids." In addition to the popcorn, visitors to KFC can also order chicken strips dipped in the chain's famous dipping sauces, including honey barbecue, "fiery buffalo," or sweet n' spicy. Fun for the entire family, perhaps, though KFC became integrally linked to tropical deforestation. The company, activists charged, was feeding Cargill soymeal to its European chickens and as a result contributed to our global climate dilemma.

Sometimes it takes daring actions to draw attention to climate change. Greenpeace has always been known as a provocative organization and, true to form, its actions against Cargill in 2006 were flamboyant. Taking to the seas in inflatable boats, activists blocked a ship called *W-One* in Amsterdam's North Sea Canal to prevent her from offloading Amazon soy from Cargill. In Britain, meanwhile, Greenpeace activists dumped almost four tons of soy in front of Cargill's headquarters, located in Surrey. Some protesters chained themselves to a gate to block company employees from entering the site.

The protests were meant to complement Greenpeace's campaign in Santarém, which had aroused the ire of the local elite. Within the town, truckers plastered bumper stickers on their vehicles reading "Fora Greenpeace" or "Greenpeace out." Reportedly, by merely displaying the bumper sticker, motorists might receive a discount on oil at a local gas station. In addition, local newspapers ran hostile editorials attacking the environmental group. Hardly deterred, Greenpeace parachuted an activist into a soy field near Santarém along the BR–163 highway. The parachute simply read, "100 percent Crime." Though the parachutist fell to earth safely it was anything but a soft landing: A small group of angry farmers rushed out to greet the Greenpeace organizer and the parachutist had to beat a hasty escape. Another planned demonstration in the rainforest had to be cancelled due to the high possibility of violence and retaliation from farmers.

Soon, however, the public mood started to shift. An editorial in the local *Gazeta de Santarém* questioned the tactics of fear and intimidation carried out by powerful farmers in the area. A soy farmer quoted in the piece surely did not help the cause of Cargill or Blairo Maggi. Santarém, the farmer said, was full of Indians and lazy people. Anti-Greenpeace bumper stickers reportedly started to decrease in number, and over 1,000 people protested in the streets of Santarém against Cargill. Escalating the pressure, Greenpeace sent its ship *Arctic Sunrise* to the Amazon. A former sealing vessel, the ship had been deployed in various campaigns on the high seas against oil pollution. From thwarting Japanese whalers intent on pursuing so-called "scientific research" to chasing illegal pirate fishermen in Mauritius, the *Arctic Sunrise* had become a cause célèbre. Perhaps most importantly, however, the ship played a pivotal role in combating climate change. In 1996, the vessel headed for Antarctica, where it documented the collapse of an ice shelf off James Ross Island. Since then, the ship returned repeatedly to the Arctic to protest oil development.

Now far from Antarctica in the Amazon, the vessel continued its mission to halt climate change. Parking in front of the Cargill port in Santarém, the ship sent teams of climbers to occupy the roof and conveyor belts of Cargill's facility. Activists displayed a banner reading "Fora Cargill"—"Cargill get out," while others sought to prevent soy from being unloaded from barges into the facility. Once activists had unfurled the banner, however, irate Cargill personnel blasted the sign down with high-powered hoses.[18] Simultaneously, the *Arctic Sunrise* attempted to occupy Cargill's dock but got rammed by a company vessel. Local police secured the Greenpeace ship, but not before an angry mob boarded the ship and painted graffiti over the sides of the boat. Using pepper spray, the police forced the crew to open the radio room while taking them into custody. As the activists were put into lockdown aboard the *Arctic Sunrise,* Cargill tug boats pushed the Greenpeace ship out of the dock and dragged it out into the Tapajós River. Greenpeace succeeded in temporarily shutting down Cargill's facility; three activists were injured during the protest and eight were arrested.

Though protests like these were small, activists were at long last beginning to build some momentum around the issue of soy and its links to deforestation in the Amazon. Would the soy industry, fearing bad publicity around the issue of climate change, start to cave and implement important environmental safeguards at long last? And what about deforestation, not just within the Amazon but in adjacent but no less sensitive ecological areas?

While some have become concerned about soy's reach into the Amazon, few are aware of the threat to the Brazilian *cerrado,* which is a shortened form of the term *campo cerrado* or "closed country." A vast plateau stretching across 500 million acres, the *cerrado* resembles Africa's safari lands. Sweeping through the central west region, the *cerrado* covers a full one-fifth of Brazilian territory and represents the country's second-largest ecosystem after the Amazon. Topographically diverse, the area alternates between forests, bushes, and grasslands. Perhaps because this scrubby mix of gnarled woodland and tropical savannah seems rough to some, the *cerrado* has not received the same kind of high-profile attention from environmental campaigns.

The *cerrado* is the world's most biologically rich savannah, with more than 10,000 species of plants, 935 species of birds, and almost 300 mammals, including endangered species like the *cerrado* fox, giant anteater, jaguar, and marsh and pampas deer. Also endangered is the maned wolf, an animal celebrated in traditional Brazilian folklore. New animals are still being discovered in the *cerrado,* including two previously unknown species of lizard said to resemble miniature ground-dwelling dragons. On a recent expedition to the *cerrado,* researchers found more than a dozen undiscovered species, including eight fish, three reptiles, an amphibian, one mammal, and one bird.

Though the *cerrado* seems remote to people in the north, it's a region that plays an important role in terms of our unfolding climate crisis. Trees in the *cerrado* are not as tall or dense as those in the Amazon, so they don't store as much carbon. Nevertheless, because the *cerrado* is three times the size of Texas, it stores its fair share.[19] That's worrying in light of the arrival of soy barons who convert the land to monoculture, which in turn liberates carbon from the soil. Talk to Brazilians about climate change, though, and at least some are likely to disagree with environmental campaigners such as Greenpeace. That's because in the span of less than a generation the *cerrado* has helped Brazil to become an agricultural giant.

The environmental cost of this meteoric economic rise is all too apparent: Today the *cerrado* is disappearing at more than twice the rate of the Amazon and every year thousands of square miles of savannah are destroyed to make room for crops. Some soy farms have grown to more than a million acres of unbroken monoculture—an area even larger than Redwoods National Park. At the current rate, *cerrado* grasslands will be entirely wiped out by 2030 unless drastic cuts are made in the amount of land cleared for agriculture.[20] Though soy is an important direct driver for deforestation in the *cerrado,* it also has an indirect impact on adjacent forest, as many cattle farms have been replaced by soy farmers who buy or rent the land. Inevitably, the cattle farmers advance into new forest area and cause more deforestation. Industrial-scale soy planting in the *cerrado* also encourages land speculators to clear adjacent forest. In a never-ending vicious cycle, poorer farmers are displaced in the process and must head toward the rainforest to practice subsistence agriculture.

What is it going to take to relieve pressure on the *cerrado* and the Amazon and slow down global emissions from deforestation? In recent years, environmentalists have poured pressure on the soy industry and they've been joined by important reformers within the Brazilian government. Indeed, before she resigned from the Lula government Marina Silva recognized the climate danger posed by the soy industry and took measures to oppose it. From her perch at the Ministry of the Environment, the former rubber tapper from the Amazon helped to burnish Lula's eco-friendly credentials by levying stiff fines on farmers responsible for cutting down too many trees. Cracking down hard, she put many farms on a blacklist, which barred them from receiving bank financing.

Predictably, the moves elicited howls of protest from farmers. "A billion people around the world are going hungry," remarked Antônio Galvan of the Rural Union in Sinop, the soy boomtown along BR–163. "Ask them if they want Brazil to stop expanding its farms." Soy baron Blairo Maggi was similarly no fan of Silva's. At a meeting in 2008 of Brazilian governors, the Mato Grosso politician showed photos taken by his state's environmental officials. The images seemed to show forest in areas that Silva claimed had been cleared. Going on the attack, Maggi declared that in some instances "newly cleared" land had been legally logged some time ago. Other areas, he added, had been consumed by natural blazes or devastated by insects. Following Maggi's push back, more than 70 percent of blacklisted farms were exonerated.[21]

However, under pressure from environmentalists, McDonald's and other major food retailers agreed to stop selling chicken fed on soy from the Amazon. Shortly afterwards, IBAMA shut down Cargill's soy processing and shipping facility in Santarém because the company lacked an environmental impact assessment. Hoping to reverse its blackened name, the Minnesota-based Cargill formed an alliance with the Nature Conservancy, agreeing to only purchase soybeans from producers who were in compliance with Brazil's forestry code, which stated that 80 percent of the forest in growing areas had to be maintained. Under a 2006 agreement, the Brazilian Vegetable Oils Industry Association which included large firms such as Cargill pledged to uphold a two-year moratorium on buying soy grown on newly deforested land in the Amazon—though unfortunately the agreement did not cover soy from the *cerrado.* In accordance with the agreement, the soy industry vowed to work with government agencies and NGOs to institute a monitoring system and safeguards to ensure that soy was not sourced from newly deforested areas. Greenpeace, the implacable foe of the soy industry, announced that soy harvested in 2008 within the Brazilian Amazon had not come from newly deforested areas. "In other words," the group concluded, "the moratorium is doing its job and halting soya related forest destruction, despite the pressure from rising soya prices."[22]

Even Blairo Maggi of all people has undergone a conversion of sorts: The soy baron recently proposed a sustainable agribusiness model based on carbon credit trading that would provide monetary value to standing Amazon forest while reconciling economic and environmental objectives. The soy magnate has now apparently found religion, though the change is not entirely based on altruism. Realizing that ecological correctness can drive up soy's value, Maggi envisions a segmented market in which high-priced soy labeled as "Eco OK" would be directed to green-minded buyers in Europe and the United States. The other soy would be low-value and nongreen and would be exported to poorer countries. Meanwhile, Maggi seeks to intensify production on previously cleared land rather than mindlessly erase more of the *cerrado*. Surprisingly perhaps, Mato Grosso's environmental standards are now among the stiffest in the world, and authorities are actually enforcing those standards. For example, growers are obliged to avoid planting on hilltops and must dedicate 20 percent of *cerrado* holdings and 80 percent of rainforest holdings to natural vegetation. If farmers fail to comply, they must pay stiff fines.[23]

Despite these positive developments some environmentalists wonder whether a one-year extension of the moratorium will be long enough to build the necessary tools to ensure that soy production will not encourage more deforestation. In the battle to control our climate destiny and soy-related deforestation, much will depend on Brazil's Ministry of the Environment, which has committed itself to speeding up efforts for the registration and mapping of rural land in the Amazon. Once the task has been completed the authorities will be able to determine which lands are to be used for farming and which are off limits for deforestation.

In the meantime global demand for soy only stands to increase, perhaps even as much as 60 percent, to 300 million tons by 2020. Driven by population growth and the increase of per capita income, the world has a seemingly insatiable desire for the commodity.[24] As long as consumers continue to purchase meat from soy-fed livestock and large banks and financial institutions continue to finance the soy industry, tycoons like Maggi will enjoy a thriving business.

Given the obstacles, then, what will it take to stop deforestation and arrest climate change? While putting environmental safeguards in place to control the soy industry and getting large international financial institutions to cut their loans to corporate agribusiness could make a difference, it will be difficult to restrain powerful economic forces in Brazil that are encroaching on the rainforest. Brazil might be persuaded to protect the Amazon as an insurance policy against more severe global warming, but the South American giant will surely be looking for lucrative economic concessions in exchange. Indeed, tropical countries will need between $10 and $40 billion per year in incentives to prevent them from turning their forests over to agriculture and other industries.[25]

President Lula rightly declares that rich nations should bear most of the cost of fighting global warming as they had been polluting the planet for centuries and consumed most of the Earth's resources. Speaking at an international conference on climate change and deforestation in Brazil, he remarked, "How can we ask the poor countries to take on the sacrifices the others didn't take on?"[26] Lula has taken to his new role of Third World defender with gusto, declaring that Europe has no right to make policy suggestions concerning the Amazon. While Europe has only 0.3 percent of its original forests, he proclaims, Brazil has at least 69 percent of its virgin forest still standing. "No one throughout the world has the moral grounds to talk about Brazil's environment," Lula said. "Before you begin to talk about Brazil, look at your own map."[27]

Here's the crucial question though: Who will pay to stop deforestation of the rainforest and how much money will be provided? To date, these are very thorny and unresolved questions. One thing's for sure though: If we are ever going to make a dent in tropical deforestation it is going to require more concerted action from the Global North. Some wealthy nations have proposed a market-based system of "carbon credits" to solve the deforestation dilemma. Initially introduced as part of the Kyoto Protocol, carbon trading permits countries whose emissions fall under the emissions cap (the allowed level of carbon dioxide equivalent emissions per year) to sell carbon credits to nations that surpass their own caps. Supporters of the scheme envision a novel global investment market based on emissions trading: Companies and countries would take advantage of incentives to invest in developing world projects in exchange for highly prized carbon credits.

If we're ever going to move forward with meaningful negotiations on deforestation the United States will have to play more of a critical role. Unfortunately, the United States has done its utmost to torpedo, slow down, and scuttle deforestation programs. At a 2005 climate change summit held in Montreal, Canada, when environmentalists pushed a proposal for compensating the Global South for preserving its forests, the United States opposed the notion. It was only under duress that the United States agreed to study the proposal. Two years later, however, during a United Nations climate change summit held in Bali, Indonesia, it was again the United States that sought to avert change. When environmentalists asked the United States to lead on climate negotiations, Washington refused and even sought to block passage of a climate agreement. Incensed, campaigners said that if the United States was not willing to participate then it should at least move out of the way and allow negotiations to proceed. Finally, the United States backed down.

Despite U.S. undermining of climate negotiations, Bali was an important milestone in addressing deforestation. As a result of the meeting, participants recognized the critical role played by tropical forests in regulating climate. Indeed, in an effort to reduce emissions from deforestation, nations sought to extend the international market in carbon credits. The basic idea is that rich nations that are unable or unwilling to decrease their carbon emissions could pay developing nations to slow deforestation. Traders have agreed to purchase and sell credits within a voluntary market, but no government can legally issue credits without the existence of an overall framework. Each U.N.-backed carbon credit would represent a ton of carbon dioxide locked away by a forest as it grows. Boosters say that the scheme could generate billions of dollars a year for developing nations that tackle deforestation. The funding could be channeled into conservation efforts designed to mitigate climate change, thus creating a "virtuous circle."

Known as Reducing Emissions from Deforestation and Forest Degradation, the REDD program, the initiative is being piloted in countries worldwide. Though REDD can take many forms, the central notion is that businesses or governments in the Global North compensate poor countries for preserving their forests, either by paying into a fund or by buying credits on carbon markets. After the Kyoto Protocol expires in 2012, REDD is expected to play a critical role. REDD's backers claim the program will provide the same benefits as reducing emissions from tailpipes and smokestacks while simultaneously helping poor people and protecting the rainforest. In a political sense, REDD could focus fresh attention on the plight of indigenous people. "For decades, capitalists, socialists, private companies, governments, and local operators have blasted into tropical communities, razed forests, and moved on with little concern for the fact that they denuded the land," says one environmental expert. REDD, however, could help to put a microscope on these issues.[28] In the wake of the Bali summit some money began to pour into the REDD program, with Norway agreeing to commit about $500 million annually to rainforest conservation. Some hope that a December 2009 United Nations climate conference held in Copenhagen will result in substantial funding for the initiative.

The fact that the world is addressing deforestation is a positive development, but as they say, the devil is in the details. In the tropics many wonder what kind of framework REDD will ultimately take. A nationalist, Brazil's Lula says that a new forest carbon market could undermine his country's sovereignty over the Amazon and local resources. With a stock market in Chicago or London defining what kind of land use should predominate in the rainforest, he fears Brazil could lose its flexibility. Instead of pushing for a market-based approach to REDD in which credits generated from forest conservation would be traded between nations, Brazil wants a giant fund financed by the Global North.

Brazil says that REDD donors should not be eligible for carbon credits, which could be used to meet emission reduction obligations under climate treaties. Lula's scheme, then, is more akin to development aid rather than carbon-based market offsets. The Brazilian president has set up a $21 billion Amazon Fund that he seeks to finance through donations from industrialized countries, individuals, and private companies. Ambitiously, the fund seeks to reduce deforestation in the rainforest by 70 percent within ten years. Norway has committed up to one billion dollars to the Amazon Fund by 2015, depending on Brazil's success in halting deforestation. Yet, it's unclear how this progress might be monitored.

In light of affluent countries' historic responsibility for worsening global warming, it's understandable that Brazil wouldn't want to let the United States and Europe off the hook. Many questions, however, remain about REDD and who might benefit from an Amazonian windfall. Some believe that for REDD to be effective at least some funding will have to be directed toward the agents of deforestation. Without a system of incentives in place, they say, there is little reason to believe that agribusiness would stop destroying forests.[29] Soy magnates see the potential economic benefits of current climate negotiations, and even Blairo Maggi now supports a so-called sustainable agribusiness model based on carbon credit trading. The scheme would provide monetary value to standing Amazon forests while reconciling economic and environmental objectives.

It would be a horrible travesty if REDD wound up bailing out the likes of Maggi without benefiting poor people. As the discussion over REDD intensifies, some Indians wonder what climate negotiations mean for them. That is, if they could get access to high-level meetings: Indigenous groups complain that they have been left out of international negotiations and they want to know who is going to control the carbon trade scheme. Indians also wonder how such schemes might affect their use and rights to the forest. Indeed, land tenure could become a key issue in the carbon-trading debate as Indians grow concerned that governments or private sector stakeholders will take advantage of carbon trading schemes to get control over indigenous territory. If REDD makes forests more valuable, then emerging "carbon speculators" could forcibly displace indigenous peoples. If they want to avert such a dire scenario the authorities are going to have to undertake massive property rights reform, clarify land titles, and halt land grabbing. Meanwhile, international negotiators should help to avert a social calamity by incorporating protections for indigenous peoples in future discussions of REDD.

Unfortunately the United States, Canada, Australia, and New Zealand have blocked such moves, a fact that has raised the ire of many Indians. At U.N. climate conferences some indigenous groups have protested with placards reading, "No Rights, No REDD!"[30] "The concept of carbon markets is driven by an economic vision," says one indigenous activist from the Brazilian Amazon. "The indigenous vision of environmental issues is based upon spiritual thinking. Indigenous peoples respect Mother Earth. It's the governments and corporations that are responsible for environmental destruction," he added.

Then there's the other sticking issue of how the funding for the new carbon schemes will work. In light of its awful environmental track record you'd think that the World Bank would be the last institution the Global North would turn to. And yet, the World Bank launched a $160 million carbon fund in Bali. Financed by developed nations, the fund was designed to slow deforestation and harmful emissions. Some have been dismayed about the nature of REDD climate-change payments aimed at helping poor countries. For campaigners it is distasteful enough that the British government would opt to shuttle its financing through the World Bank's multilateral fund. But when activists learned that the authorities would provide money not in the form of aid but loans they went through the roof. "It is outrageous that the UK is prepared to make poor countries even more heavily indebted trying to combat a problem they did not cause," remarked one.[31]

Word of the large financial institution's involvement immediately raised suspicions among critics, who argue that the World Bank has historically done much to hasten deforestation.[32] In a joint statement issued during recent climate negotiations in Poznan, Poland, more than 140 groups working to halt climate change argued that the World Bank should not be involved in any carbon scheme, and instead called for climate funds to be managed under the United Nations Framework Convention on Climate Change. While the Global South urgently needed billions of dollars to cope with climate change and to build low carbon economies, they argued, funding would have to come through the United Nations climate convention, in which all parties had an equal say. "It is simply outrageous for climate financing to be given to southern countries in the form of loans," said one environmental campaigner. The activists have a point: For peoples of the Global South climate change and poverty form a vicious cycle. Part of the reason poor countries are having such a difficult time confronting climate change in the first place is that they must pay debts to northern countries. The World Bank and rich countries, which bear great responsibility for the climate crisis, unjustly want the Global South to assume the cost of dealing with its effects and thus add to their debt burden.[33]

Despite these problems, REDD could make a real difference in addressing climate change if done correctly and if forest dwellers wind up benefiting from the plan. Indeed, some indigenous peoples, instead of protesting the scheme, are seeking to take advantage of the upcoming REDD windfall. In the Juma Sustainable Development Reserve, encompassing 1.4 million acres of Amazonian rainforest, foreign businesses or governments may now purchase offset credits on the voluntary carbon market. The funds go toward protecting the reserve's thriving ecosystem, in part by compensating Indian families for protecting the jungle. Here's how it works: Each family receives a monthly stipend and local villages get solar panels, computers, and funds for education and clinics. Meanwhile, deforestation in the area is monitored by satellite to insure compliance with the scheme. If deforestation takes place or there is rampant destruction, the family that owns the land must drop out of the program and the village is put on warning. Though the project is still in its infancy it could prevent the release of 190 million tons of carbon between now and 2050 if it proves successful.

As we near climate negotiations at Copenhagen the political challenges facing the planet are stark. Indeed, even REDD's strongest boosters readily concede that fulfilling all the hopes within the policy while avoiding potential pitfalls will be a risky proposition. There's a lot riding on REDD: Environmentalists and Indians want the scheme to help improve governance, promote sustainable development, and mitigate climate change—all at the same time. It's certainly a tall order, and some may wonder whether REDD is politically viable. Despite the many obstacles, parties to the REDD process including governments, NGOs, the private sector, and indigenous peoples have made a lot of progress on the issues and challenges. Whatever its flaws or shortcomings, REDD is the only game in town right now that makes preserving forests more economically valuable than cutting them down. "REDD may be our last, best hope of saving the tropical forests, which are so essential to the future health of our planet," writes one environmentalist.[34]

Chapter Six

DEFORESTATION DILEMMA

E ven as Brazil wrestles to save its rainforests, other South American nations are struggling to cope with increasingly severe deforestation problems. Peru has historically had one of the lowest yearly deforestation rates in the Amazon, though forest loss has been on the upswing in recent years. Forests cover 60 percent of Peruvian national territory, and deforestation razes more than 250,000 hectares of trees every year. That's much less than Brazil, which cuts down twelve million hectares annually, but still destructive enough to make conservation a top priority.[1] What happens environmentally in Peru is important—the country has the second largest swathe of Amazon after Brazil, covering 16 percent of the overall rainforest. The Amazon River has its origin in Peru, and the nation has some of the largest and most biologically diverse forests on Earth. Currently Peru accounts for a tiny percentage of worldwide carbon emissions—less than 1 percent. Could deforestation drive up Peru's contribution, spelling future danger for world-wide climate change? It's a question of vital importance for the Global North, which has been buying up tropical commodities from Peru and promoting an unsustainable free-trade model through its large financial institutions, furthering deforestation and hastening the day of our climate reckoning.

In Peru subsistence farming is driving most deforestation, but as in Brazil other important extractive industries push rainforest settlement forward. In recent years logging, mining, agricultural expansion, cattle ranching, oil extraction, and even illegal coca farming have figured prominently. For years the United States and local authorities have sought to eradicate coca production by military means. This in turn has led the cartels to strike back and to eliminate anyone who dares to get in their path. Caught in the crossfire are Indian farmers, who are pushed into coca cultivation as a result of dire poverty. While farming coca in the tropical jungle is a highly risky enterprise, for many it's simply the only way to support their families.

Like Brazil, Peru has long been characterized by intractable social inequality—inequality that now stands to have an environmental impact upon us all. Think about the cocaine trade and the first image that may come to mind is of egomaniacal, bloodthirsty *capos* like Pablo Escobar. But while Escobar and his ilk have enriched themselves off illicit profits, poor farmers don't share nearly as much in the proceeds. Fleeing poverty in the Peruvian Andes, many *campesino* peasants head to the jungle where they chop down the rainforest to plant coca leaf. The newcomers occupy the rainforest illegally; farm plots for no more than five years, until the soil is depleted; and then move on to other lands to start anew.

Drug users in the United States and Europe may not consider the social costs of the drug trade. Nor, for that matter, will it occur to them that coca production has resulted in tropical deforestation and a worsening of climate change. If people of the Global North want to do something to conserve forests and help ameliorate climate change, they should work to make coca production less lucrative. That's because over the past few decades, millions of acres of Andean forest have been lost to plant coca and opium poppies. During the 1980s, coca leaf became the largest cultivated crop in the Peruvian Amazon. Experts found that coca resulted in the deforestation of more than a million acres of rainforest, an area roughly twice the size of Rhode Island. It is a doubly destructive crop, because workers must also deforest adjacent areas to create plane landing strips and to grow subsistence crops of corn, bananas, and yucca.

By the 1990s the cocaine trade came to be centered in the Upper Huallaga River Valley of Peru, up above the Amazon. Spurred on by sky-high U.S. and European demand, planters plowed through fragile cloud forest, known as the "eyebrow of the jungle" for its altitude and lushness. The planters concentrated their efforts on tropical slopes ranging from 3,000 to 6,000 feet, where soil contained a high content of alkaloid, an active ingredient in cocaine.

The Upper Huallaga River Valley is located within the Peruvian Yungas region, comprised of sub-tropical montane deciduous and evergreen forests flanking the eastern slopes and central valleys of the Andes. The vegetation here is extremely diverse, with more than 3,000 species of flora and at least 200 species of orchids. Endangered species such as jaguar, ocelot, and spectacled bear roam the treacherous slopes. In addition to providing a welcome habitat for unique Andean species, cloud forests in the area perform a vital environmental function: carbon sequestration. Cloud forests store just as much carbon as lowland forests, though the carbon is more often stored underground, in roots, soil, and litter layers.

Though cloud forests account for just a small portion of the Earth, they contain a disproportionate share of the planet's plant and animal species. Cloud forests like the ones around the Yungas also play a positive role vis-à-vis climate change as they harbor bamboo, a plant that acts as a valuable carbon sink. A gigantic grass with hollow, woody stems, bamboo has many versatile uses—it can be made into panels, boards, flooring, roofing, pulp, and paper, as well as fabrics and cloth. Bamboo miraculously grows up to one foot every day and can be harvested every five years on average. Most trees, by contrast, take much longer to grow—from 40 to 120 years. This quick growth means that bamboo can sequester more carbon from the air than other, slower-growing species.[2]

If we want to preserve our climate equilibrium we're going to have to do a much better job of protecting cloud forests. The clouds are formed when warm winds are forced upward by the mountains, then cool and condense. These cloud forests, as well as the unique plants and animals that dwell in them, depend on this prolonged cloud immersion. But now, scientists say, deforestation may be causing weather to change in the lush cloud forest. Here's how it works: As trees below the cloud forests are replaced by farms, roads, and settlements, less moisture evaporates from soil and plants. This in turn reduces clouds around forested peaks lying miles away. Scientists say that if current rates of deforestation continue, cloud forests will be forced upward. As they go higher and higher, cloud forests will decrease in area, become increasingly fragmented, or even disappear.

Sadly, as long as there's a voracious demand for cocaine in the rich countries of the north, cloud forests will suffer. When the authorities clamped down on coca production in Peru during the 1990s, the planters simply picked up and went to Colombia, where coca production sky-rocketed fourfold. More than three million acres of rainforest—an area larger than Yellowstone National Park—were cut down to cultivate opium poppy and coca leaf.[3] As in Peru, coca cultivation quickly spread into the cloud forest, thus provoking deforestation and exacerbating climate change. It wasn't long before whole coca-growing villages sprung up in-side the Macarena Reserve, a biodiverse area containing cloud forest.

Coca production in Colombia's cloud forests put intriguing animal species in jeopardy, and some cloud forest species may be wiped out be-fore scientists can ascertain much information about them. Take, for ex-ample, the gorgeted puffleg hummingbird, which was only discovered in 2005. It displays flamboyant colors—males have an iridescent patch on the throat and white tufts above the legs—and conservationists warn that unless the hummingbird's habitat is protected from deforestation and coca production the animal could be wiped out. The habitat of the Cloud-forest Pygmy-Owl has also been diminished by coca plantations, and un-fortunately, it's difficult for us to know about the owl's status as the animal is notoriously difficult to observe.

Like Peruvian cloud forests, Colombia's cloud forests contain stands of bamboo. In addition to fulfilling a vital environmental function through the absorption of carbon, bamboo is used to construct houses and furniture. Furthermore, local inhabitants use bamboo to make hand-icrafts and musical instruments such as Andean pan pipes and flutes. Sev-eral important mammals feed on bamboo, including the spectacled bear, which eats the plant's young shoots. Classified as vulnerable by the In-ternational Union for Conservation of Nature, the spectacled bear is black and may have yellow markings around the eyes, hence the animal's com-mon name. One of the most emblematic mammals of the tropical Andes, the animal is the only species of bear in South America. In recent years, the spectacled bear has seen a reduction in its native cloud forest habitat as coca production has expanded. The United Nations Environment Program reports that deforestation has affected the bear's migration routes and the animal has moved into farms in search of food. This in turn has led to conflict with local settlers.[4]

The cocaine-fueled destruction of the cloud forest and resulting environmental impact on climate is not a point that has been lost on Colombian policymakers. Speaking to a conference of police officers in Belfast, Northern Ireland, Colombian vice president Francisco Santos Calderón decried land clearing and deforestation resulting from coca production. Indignantly, Santos declared that Brits using cocaine should be more conscious of the environmental impact of their drug habit. "Colombia has lost more than two million hectares of rainforest in the last 15 years to plant coca," he said. "If you snort a gram of cocaine, you are destroying 4 meters square of rainforest and that rainforest is not just Colombian—it belongs to all of us who live on this planet, so we should all be worried about it. For somebody who drives a hybrid, who recycles, who is worried about global warming—to tell him that that night of partying will destroy 4 meters square of rainforest might lead him to make another decision," Santos added.[5]

While Santos makes a valid point about the ecological costs of cocaine production, U.S.-backed anti-drug efforts in the Andes have failed to halt the coca trade or, for that matter, to eradicate environmental problems stemming from this illicit industry. Indeed, when the Colombian authorities clamped down on coca growers, production was simply driven back into Peru. Today, the Yungas finds itself in almost a critically endangered condition, in large part because of the coca trade. In the absence of viable alternatives, coca production looks set to continue in the Upper Huallaga River Valley. It's the law of simple economics: Farmers harvest coca four times a year and sell it for $3.30 per kilo. Coffee and cacao, meanwhile, yield only one crop a year and sell for just $1.25 to $1.50 a kilo.[6] According to coca growers, U.S.-sponsored programs that pay farmers to grow alternative crops such as vegetables have not worked. The United Nations has stated that Peru needs significantly more international financial aid for crop substitution in order to halt the coca trade.

When you descend from the cloud forests of the Yungas into the jungle, the tragic consequences of the coca trade become all too apparent: Vast swathes of lunar-like desert peek out from the midst of virgin rainforest. Currently, coca is responsible for a full 25 percent of deforestation in the Peruvian Amazon.[7] Consumers of the Global North are complicit in this destruction, since most of Peru's cocaine gets exported to Europe. As Colombian vice president Santos noted, the government's own drug czar has estimated that for each line of Peruvian cocaine that gets inhaled in London or Paris a full three cubic meters of virgin jungle gets ripped down. It's an alarming statistic but could be essentially true. In Peru, one hectare of land yields 1,000 kilos of cocaine per harvest on average. And to produce one kilo of high grade cocaine one must harvest 360 kilos of coca leaf simply on that one hectare of land. Multiply the coca production on that one plot exponentially and you wind up with one large deforestation problem with implications for global warming.

While coca is certainly a significant component of deforestation, we need to consider another important driver: logging. Unfortunately, in Peru both interests are tied together in an insidious web: The loggers launder drug money and much of the mahogany harvesting occurs in the same regions where coca is cultivated under the control of drug cartels.[8] When people denounce the logging trade in Peru or public officials refuse to collaborate with the logging "mafias," they receive warnings via third parties that they should take care or risk everything. It's a brutal industry that has received the patronage of the Global North for far too long—mahogany consumers have been complicit in the deforestation dilemma.

Of particular concern is the practice of so-called selective logging, which is suspected of playing a large role within the local Peruvian economy. In contrast to clear-cutting, which chops down vast swathes of forest, leaving little behind save wood debris and a lunar landscape, selective logging removes a few trees and theoretically leaves the rest intact. If this all sounds sustainable, think again: For every tree that gets chopped down up to thirty additional trees can be seriously damaged. That's because when trees are felled the vines that wind through them pull down nearby trees in a kind of avalanche effect. Furthermore, selective logging employs tractors and skidders that rip up the soil and forest floor. As loggers plow through the forest they build makeshift dirt roads that encourage colonization of the rainforest by land-hungry settlers.

If that was not enough reason to doubt the long-term viability of selective logging, consider also that this destructive practice leaves gaps in the forest canopy that permit blazing sunlight to heat up the moist rainforest ground clutter. That's a problem because normally the jungle is a very shady place and only 1 to 2 percent of the sun's rays reach the forest floor. Selective logging has changed all that—now moisture that otherwise would have been retained by the jungle evaporates. It all makes for a very "inflammatory" situation as the forest floor gets turned into a tinder box. Selective logging also gives rise to so-called "surface" fires that burn the forest floor, destroying shrubs and floor litter. Though surface fires are destructive simply on their own, the blazes may be more harmful for their long-term effects as they dry out the forest and pave the way for future conflagrations. Over time, the forest may get so dried out that it turns into savannah or degraded scrub.

The Peruvian Amazon contains more than 300 species of trees, but what the loggers are really after is the so-called "red gold": pricey *Swietenia macrophylla* or big-leaf mahogany, with a reddish sheen. A tree that takes seventy-five years to mature, mahogany majestically peeks over the jungle canopy and can grow 120 feet high. Because of its beauty and durability, mahogany is highly sought after and is used as sawn wood, plywood, and veneer. It is also used to make high-class furniture, yachts, and even musical instruments. Money grows on this tree: One mahogany can produce more than $100,000 worth of high quality furniture. Once widely available throughout Central and South America, mahogany has become so overharvested that it is now only available in significant quantities in the Amazon. After felling the trees with chainsaws, loggers saw the wood into planks and lash them to a makeshift raft. Eventually the wood winds up in the jungle town of Iquitos and gets exported abroad. A flashpoint for future environmental woes, the removal of mahogany opens the path for more logging and the agricultural degradation of the rainforest. Even within the rainforest clumps where they grow, let alone outside these patches, mahogany density is not very high. This means that loggers must construct long roads into the jungle if they want to get at the trees.

So, just how concerned should we be about the climatic effects of selective logging? Scientists have sounded the alarm bell. They say that logging is contributing to global warming and more intense El Niños since trees absorb carbon dioxide and exchange it for oxygen. When forest fires strike, they release that carbon dioxide. Moreover, when loggers remove a tree trunk they leave the crown, wood debris, and vines behind to decompose. This releases carbon dioxide gas into the atmosphere, adding to our climate woes. In 2005, one researcher calculated that selective logging in the Amazon was responsible for the emission of one hundred million tons of carbon dioxide into the atmosphere, which came on top of the already huge carbon dioxide emissions attributed to traditional deforestation—another 400 million tons.[9] More recently, the Nature Conservancy has estimated that selective logging in the tropics is responsible for a full 30 percent of emissions from deforestation.[10] Cutting the extracted tropical wood in sawmills adds yet another unwanted environmental dimension. Many sawmills only have an efficiency level of approximately 30 to 40 percent, which means that large amounts of sawdust and scrap decompose into atmospheric carbon dioxide.

A thorny climatic problem to be sure, yet solving the deforestation dilemma may prove challenging from a social standpoint. Like coca growers, loggers are often pushed into their trade as a result of desperate material circumstances. In Peru, where one out of five people live in poverty, mahogany is a tempting target as the tree can fetch hundreds of dollars at the sawmill. Not that the loggers themselves have gotten rich from the proceeds. Indeed, grizzled loggers make only about $7 a day and spend up to six months deep in the jungle in search of valuable trees. Over time the men, most of whom are Indian or *mestizo*, fall into a vicious cycle of debt that ties them to their bosses almost for life.

Here's how it works: Middlemen contract and equip the loggers, providing the men with everything they need to survive in the jungle and cut down trees. The workers in turn are obliged to deliver the cut timber. But when the loggers return the timber to the boss, he argues that the wood is no good and reduces payment arbitrarily. At the same time, the workers fall drastically into debt because the bosses charge them exorbitant rates for supplies out in the jungle. If the men want a soda, for example, they must pay the equivalent of one day's salary. Even years later, the loggers are often still indebted to the bosses. In Lima and the poor Iquitos neighborhood of Belén, it's not uncommon for Indian men to literally give their daughters away to the logging bosses as payment for old debts, and the girls must work as sex slaves at the unscrupulous whim of the bosses.[11] Escape for the loggers is an iffy proposition: The camps are isolated and the bosses strip the workers of their documentation while physically threatening the men.[12]

Some loggers are certainly aware of the environmental downside of their work but feel they have little choice but to continue. When scientists try to confiscate the loggers' chainsaws, the men weep and cry out that they must feed their families.[13] Recently the United Nation International Labor Organization estimated that in Peru more than 50,000 loggers working in slave-like conditions were toiling away in 1,500 logging camps located throughout the country. Living in the jungle, the men survive as best they can, taking shelter in thatched huts and foraging for food. Feeding on spider monkeys and wild pigs, the men do their best to survive under grueling conditions. After toiling in the heat, some loggers age prematurely and look much older than their true age.

As the loggers move into the jungle, climate change and deforestation spur the spread of malaria in the Peruvian Amazon. Unseasonal rain is changing the patterns of mosquito development, leaving puddles of lethal larvae in regions where malaria had been nonexistent formerly. Deforestation, meanwhile, forces mosquitoes to move to new areas and spreads the disease to regions where people are unfamiliar with malaria and lack the means to acquire mosquito nets or preventative medicine. In frontier areas that have been stripped of trees, towns and roads are far more likely to harbor malaria-infected mosquitoes. Researchers even found that the mosquito biting rate in deforested areas was 300 times greater than in pristine rainforest.[14] In the logging camps workers are tormented by mosquitoes, and by the time they make their way out of the rainforest the men are often suffering with malaria or other tropical infections. Some eventually die as a result of disease. After coming into contact with loggers, some non-logging Indians die from malaria.

What will it take to stop the mahogany trade and halt carbon emissions stemming from deforestation? Through their demand for expensive wood, Americans and Europeans are exacerbating climate change. Peruvian mahogany has been used to construct everything from writing desks worth nearly $10,000 to acoustic guitars valued at $3,700 to the decks of suburban homes, and even coffins. Because of the lucrative profits involved, authorities have been reluctant to clamp down on the mahogany trade. During the Bush era in the United States, officials moved to oppose new mahogany protections aimed at ensuring sustainability. Small wonder, given that 60 percent of the mahogany from Latin America went to the United States.[15]

Peru, the largest mahogany exporter in the world, sent 45,000 cubic meters of mahogany to the United States in 2002. To put it in perspective: that's the equivalent of 50,000 trees. Though Peru set aside areas of formal logging concessions, 90 percent of the wood shipped to the United States was logged illegally. In the rainforest, logging has been everything but clandestine. The camps are known to all and the logging work itself is noisy. Day after day, flat barges float down the river, loaded with timber. You might think that stopping logging would be an easy task because the business is so visible. Corruption, however, has made crackdowns unlikely. During the Bush years, the rainforest took a heavy toll as a result of the mahogany trade. Before winding up in American homes, the mahogany passed through a chain of corrupt officials and exporters. They in turn purchased or forged documents through customs agents as well as importers and timber yards that could not be sure of the wood's origin. The whole issue was compounded by Peru's National Institute for Natural Resources (INRENA), in charge of regulating flora and fauna, which was blamed for much of the corruption plaguing the timber trade. In American furniture stores, meanwhile, retailers continued to sell elegant carved dining tables with the unmistakable rippled grain of mahogany. No one bothered to ask where the wood came from.[16]

Concerned that deforestation could veer out of control with dismal consequences for climate change, Greenpeace activists undertook a daring move: In 2002 they tried to climb up onto a commercial ship, the *APL Jade*, off the Florida coast. The boat had seventy tons of mahogany on board which had reportedly been illegally cut from the Amazon. The activists attempted to hang a banner on the ship reading "President Bush: Stop Illegal Logging." Approaching the *APL Jade* in a tiny open boat, the demonstrators identified themselves and ceased efforts to hang the banner once the ship's crew stopped them. In response to this innocuous protest the Bush administration came down hard, accusing Greenpeace of terrorism. In a case peppered with references to 9/11, the Justice Department charged Greenpeace under an obscure nineteenth-century law that was originally intended to deter vagrants jumping aboard ships, like something out of *Pirates of the Caribbean*. In its original indictment against Greenpeace, the Department of Justice also claimed that the cargo aboard the *APL Jade* was not illegal. Later, however, the government abandoned its claim that Greenpeace was wrong about contraband cargo aboard the ship, which was ultimately unloaded in the port of Charleston, South Carolina.

It wasn't until two years later that a federal judge threw out the criminal charges and dismissed the case. Environmentalists, meanwhile, continued their legal battle against the government, suing the Department of Homeland Security, the Department of the Interior, and the Department of Agriculture to stop turning a blind eye to the illegal mahogany trade and to enforce the law. Under the U.S. Endangered Species Act and international conventions, they argued, importing mahogany into the country was illegal. They also appealed to U.S. companies to stop buying mahogany from Peru until the Andean nation could organize a lawful and sustainable forestry system.

After years of being asleep at the wheel in terms of the timber trade, there are now encouraging signs that the United States is finally taking the illegal mahogany issue seriously. Under a new revision of the Lacey Act, trade which aids and abets the illegal lumber trade is prohibited. A law with ambitious scope, the statute actually makes it illegal to transport or sell wood that violates the laws of indigenous tribes. Moreover, U.S. importers must declare the country of origin and species for all imported timber and furniture. The law has already had an extraordinary impact since up until now companies involved in the timber trade had been acting under a "don't ask, don't tell" policy: once illicit wood left South America it was fair game on the open market.

Hopefully the Obama administration—which takes global warming more seriously than the administration of George W. Bush—will actively enforce the new law so as to take care of the demand side of the illegal timber trade. But what about the supply side and could the current Alan García regime take a different tack, protecting the rainforest so as to head off the worst ravages of climate change? While environmentalists recently charged that up to 90 percent of Peru's exported mahogany was still illegal, the Peruvian government flatly rejects such arguments. President García claims his government has "turned over a new leaf" and that authorities strictly regulate the export of all woods through logging concessions with no illegal mahogany slipping through to the United States. But even if the number of illegal exports is going down as claimed, that doesn't mean that illegal logging as such has gone out of fashion. Indeed, García himself concedes that the practice continues, while indigenous leaders point out that few forest rangers enforce the law in the remote areas where mahogany trees remain.

García's much-touted logging concessions may be part of the problem. Satellite images reveal a great deal of "leakage" outside concession areas into neighboring forests, meaning that loggers or concession holders simply filter over the established borderline and start felling trees elsewhere. If the loggers cut down mahogany illegally it's very difficult to monitor or trace the origin of the trees.

Try to challenge the power of the loggers and you can get into trouble. A community leader and Brazil nut collector or *castañero* in the Madre de Dios region of the Amazon reported a truckload of illegal mahogany to the police and INRENA. After making his report of the illegal shipments, the Brazil nut collector was shot repeatedly and killed for defying the loggers. The assailant carried out his attack just feet away from the local police station, suggesting that the loggers have become more brazen and do not have much to fear from local law enforcement.[17] A recently created Ministry of the Environment, meanwhile, only has sixty-one enforcement agents at its disposal. "It's a joke," remarks the new environment minister, Antonio Brack, who is seeking money to fund a larger force of 3,000 agents.[18]

Hopefully Peru won't turn into a deforestation disaster like the Brazilian Amazon. Yet coca and logging have become intractable environmental problems and could worsen our climate dilemma in the coming years. To what extent will the Peruvian authorities be able to tackle climate change on their own? Given the pattern of recent years, it's hard to be very optimistic about the future. Nevertheless Jorge del Castillo, Peru's prime minister, has called global warming a "serious threat" to the planet. What's more, President García recently announced the creation of a new environmental ministry.

Though the president's creation of the ministry represents progress, some believe that the government's move to create the new agency had more to do with satisfying economic exigencies than protecting the local ecology. Indeed, Peru was reportedly required to create the state entity in order to receive a crucial Inter-American Development Bank loan. Government officials frankly admit that it's going to be an uphill battle for the new ministry. One young environmental engineer at the ministry told me that she and her colleagues have even held showings of Al Gore's film about climate change, *An Inconvenient Truth*, to educate journalists and members of Congress. Unfortunately, the Ministry of Economy and Finance doesn't see the environment as a priority. "We're trying to convince them that the issue of the environment is interesting and important for the development of our country. But they still see it as a green issue," she says.[19]

To head up the new Ministry of the Environment García tapped Antonio Brack, a well-known environmental scientist. Born in the Amazon, Brack grew up with Yanesha Indians and went to grammar school in a modest schoolhouse that had only one teacher. One of eight children, his family was poverty stricken and at times there wasn't enough to eat. It wasn't until he reached the age of eleven that Brack actually put on a pair of shoes. Fortunately, the family grew more prosperous when they started to grow high-quality coffee. Though his father had only a fifth-grade education, Brack became intrigued by science and started to purchase books. He became more interested in nature after observing a local Indian doctor named Doña Narcisa who treated her patients with medicinal plants.[20] When he was twenty-one, Brack published a scientific article about the Andean spectacled bear. A precocious youth, he published his first book a scant four years later. Pursuing his scientific studies in Europe, he supported himself by working as a barber and a newspaper deliveryman. After returning to Peru, Brack oversaw a special program within the Ministry of Agriculture designed to conserve the Andean vicuña.

With Brack at the helm, is there hope that Peru could adopt a more progressive stance on climate change? The new minister takes deforestation seriously and recently remarked, "One of the worst problems about global warming is that mankind in the last 500 years has destroyed 50 percent of forests on the planet and that is a very serious problem indeed."[21] In an effort to roll back negative history, Brack says Peru can play a key role in offsetting carbon emissions and participate in novel carbon trade schemes. Recently it's become fashionable in Latin America to atone for environmental sins by carrying out initiatives toward so-called "zero-carbon" emissions that would reforest trees and thereby neutralize greenhouse gases produced by industry and other causes. Ambitiously, Brack says Peru can reach zero deforestation in just ten years with the proper assistance. "We are an important country with a large area of forest that has a value," Brack exclaims.[22]

But who will pay to conserve the forest? Recently the new minister announced a government plan that would pay indigenous communities 5 soles ($1.70) per hectare ($0.68 per acre) for preserving the forest. The scheme, which is scheduled to commence in 2010, would generate $18.3 million for local people.[23] When critics countered that the money was a pittance and didn't demonstrate enough seriousness on the part of the government, Brack doubled the amount paid to indigenous communities to 10 soles ($3.30) per hectare for protecting the rainforest.

García has called on world leaders to create a global reforestation fund that would be financed by a fossil fuels tax. Jumping on the green bandwagon, the president recently urged Peruvians to join him in a national reforestation campaign to reverse global warming. "We must make Peru green . . . each [of you] can do it in front of your house. Go out of your house and plant a tree," the president said. The new program, spear-headed by Peru's agriculture ministry, aims to reforest millions of trees in the Amazon. The president wants to authorize the sale of vast tracts of deforested, uncultivated land in the Amazon to private companies that pledge to pursue reforestation. "We live in an ideological world that says the Amazon cannot be touched, because it is part of the idyll of primi-tive communism," García said. Peru should not fall prey to such delu-sions, the president added, because reforestation would generate jobs and attract investment.

Peer below the president's rhetoric, however, to find the devil in the details. Because there is no land registry and many do not hold formal title to the land, natural areas can be sold off. Indeed, the draft law did not clearly define what kind of lands would be affected. That means that an ostensibly green program could wind up going horribly wrong: oppo-nents fear that a plot of land "without forest cover" could actually contain primary forest which could be destroyed once the property was sold. Under García's plan, it is unclear how land sales would be carried out, what rights purchasers would acquire, or how local residents would be affected. Hitting back at the president, one opposition legislator said it was García who had turned delusional. "We don't want our natural re-sources to be poisoned," he said. "Even the present system of forestry concessions has proved to be a failure, because it has only served to plun-der the jungle, allowing a small group of companies to benefit from the sale of illegally logged wood." Social organizations in the Amazon agreed, declaring that García was merely copying foreign models of reforestation without taking local residents into account. Incensed by the central gov-ernment, they organized street protests against García in local towns.

The distrust felt by many Amazonian people toward the García gov-ernment has important implications for future climate discussions. While Brack and others at the Ministry of the Environment may be well inten-tioned, they could be undermined by other government figures. As dis-tressing as it sounds, we're going to have to wake up to the ugly truth that even if wealthy countries come up with Brack's money to protect the

Peruvian rainforest, there's no guarantee that the funds won't get siphoned away by corrupt officials. Because of the risks, it's important for future deforestation measures to include good governance provisions so that poor local communities actually receive the funding they deserve. It's a point that hasn't been lost on environmental campaigners, who have lobbied for good governance during international climate summits.

Hopefully the international community will be able to address these concerns at upcoming climate talks in Copenhagen, Denmark. Meanwhile, some affluent nations seem to be taking their carbon debt to the Global South somewhat seriously. Brack is encouraged that Japan has agreed to loan his country $120 million to protect 212,000 square miles of Amazon rainforest over the next ten years. The minister estimates that the initiative will avoid 20 billion tons of carbon dioxide emissions. Germany has also joined in the fight by providing $7 million toward forest conservation. But Brack says Peru needs more money—approximately $25 million a year over the next ten years, to be exact—in order to conserve 54 million hectares of forest. If the Reducing Emissions from Deforestation and Forest Degradation (REDD) program does kick in then perhaps Peru will start to benefit. Under the scheme, forest land could be worth $800 or more per hectare for its carbon, depending on the threat level. Those forests at high risk for deforestation would be compensated at a greater rate than inaccessible areas that are at a low risk for development. With so much at stake, affluent nations are going to have to get the REDD deal right at Copenhagen.

One can only hope that climate justice will prevail, but a disturbing possibility is that for every one step forward Peru may take one step back. That's because the United States, far from taking a constructive role, is helping to drive an unsustainable model in the Amazon. It all boils down to the crucial question of energy politics: In recent years Washington has allied itself with the corrupt García regime in Lima in order to extract the rainforest's natural resources. It's an effort that has already antagonized indigenous peoples and stands to worsen climate change for all.

Given the negative effect that fossil fuels have on climate change, you'd think the United States would assist Peru in making its transition toward a safer and greener development model. To the contrary, however, U.S. companies are pursuing problematic mega-projects backed up by large financial institutions. Take, for example, the case of the Camisea gas field. Located in the remote lower Urubamba Valley in the southeastern Peruvian Amazon, the $1.6 billion project includes two pipelines to the Peruvian coast. Politicians have seen Camisea, with its proven reserves of 11 trillion cubic feet of natural gas and 600 million barrels of oil and liquid gas, as crucial in providing for the energy needs of Lima's eight million inhabitants.

Hardly a boon to climate change efforts, Camisea will generate almost 1.1 million tons of carbon dioxide per year and 133 million tons over the course of the project's lifetime.[24] That's greater than the emissions for all of Central and South America combined (excluding Argentina and Brazil) in the year 2000 and more than the entire African continent combined (excluding South Africa) for the same year.[25] According to Indians, the project has led to deforestation due to widening of the right of way for the pipeline and detours from the established route. Furthermore, they charge, work on the project has been carried out within the Nahau Kugapakori Reserve, an area established for the protection of Indians living in voluntary isolation from the outside world.[26] The pipelines, scientists say, cut through a biodiversity hotspot that should be "the last place on earth" one would drill for fossil fuels.[27]

At the heart of this deforestation and climate boondoggle is none other than Hunt Oil and Kellogg, Brown & Root, two U.S. companies. Hunt is a subsidiary of former vice president Dick Cheney's company Halliburton. Hunt Oil's CEO reportedly raised more than $100,000 for Bush's 2000 presidential race. One of the more controversial pieces of the Camisea project was Kellogg, Brown & Root's proposal to construct a liquefied natural gas plant that would convert natural gas into butane, diesel, jet fuel, and gasoline for export to Mexico and the United States. Indeed, boosters plan to ship half of Camisea's cheap gas to supply U.S. West Coast markets, which would undermine California's renewable energy initiatives.

If we want to control our climate destiny we must seriously examine the role of large financial institutions. As in Brazil, where it backed carbon-unfriendly industries such as cattle and soy, the World Bank has played an environmentally questionable role in Peru. Through the International Financial Corporation, the bank approved a whopping $300 million loan for the Camisea liquefied natural gas export project operated by Hunt Oil. Meanwhile, the Bush administration pushed hard for Camisea and pressured the Inter-American Development Bank (IADB) to provide funding for the project.[28]

Like the World Bank, the IADB has emerged as a key player in South America. IADB involvement in Camisea hardly comes as a surprise: The financial institution has long supported outdated, climate-worsening fossil fuels and has insisted on export-led development to the detriment of alternative, sustainable development that helps everyday people as opposed to the private sector.[29] The United States is the IADB's main shareholder, with 30 percent of voting power. When it comes to electing the president of the IADB, member countries with the most capital, like the United States, have a lot more power than poor countries. Activists say that the IADB does not promote open dialogue about its projects. One environmentalist campaigning against Camisea told me that an IADB meeting in Miami was "hyper stage managed." The activists could ask questions of the IADB president but were not allowed any follow up.[30]

The IADB has also played a role in helping to push the U.S.-Peru Free-Trade Agreement, which has hardly been a boon to the environment. The treaty, supported by the former Bush administration as well as some prominent Democratic legislators such as New York senator Chuck Schumer and even then–Illinois senator Barack Obama, has led to incredible social and political polarization in the Peruvian Amazon. Under the agreement, President García was obliged to sign decrees that made it easier for foreign developers and oil interests to buy Amazon rainforest land. But after García signed the decrees he found himself in a political pickle. That's because the decrees weren't opening up mere empty lands but areas inhabited by actual indigenous peoples. As he sold tens of millions of dollars worth of exploration rights to foreign companies, García failed to consult with local tribal leadership in the rainforest, thus setting the stage for a nasty showdown. Outraged, the Indians wanted to know under whose authority García had sold ancestral lands to foreign companies. Never fear, García declared: Under "fast track" authority granted to him by Congress, the president could make laws that would facilitate the implementation of Peru's free-trade agreement with the United States.

Peru is not a net oil exporter, but the current Alan García regime in Lima has high hopes for the sector and says petroleum investment is necessary to boost economic growth. A political scoundrel and opportunist, García is hardly an ideal politician to tackle global warming. He shocked many when he published a series of controversial editorials in the Peruvian paper *El Comercio* criticizing environmental nongovernmental organizations (or NGOs) and indigenous organizations. In the columns, García argued that the groups were doing their utmost to block Peru's successful economic development. The president's words reinforced what progressive-minded folk always suspected: that at the end of the day the authorities perceived environmentalists as communists, socialists, or even caviar-eating elitists.[31] In 2007, after arrows and abandoned camps were found in western Brazil, providing evidence that isolated, uncontacted Amazonian tribes were fleeing Peru to escape encroaching loggers, García suggested that such indigenous groups were merely an invention by pesky critics opposed to oil exploration.

Unfortunately, García's free-trade agreement with the United States stands to exacerbate global warming and environmental problems in the Amazon. Today vast stretches of rainforest land have been turned over in concessions to oil companies. The numbers are staggering: More than 70 percent of the rainforest has been allocated for oil and gas extraction, and many of the concessions overlap with protected wildlife areas and indigenous reserves. García's pro-oil policy completely undermined more progressive sectors of his government: Even as the Ministry of the Environment sought funds for the REDD scheme, the Ministry of Energy and Mines was leasing oil and gas concessions in the Amazon. While there are few regions where actual drilling has commenced, oil exploration encourages deforestation, as workers must construct heliports and camps. The main concern, however, is road construction, which leads to landed settlement and more illegal logging. In addition to affecting climate change through simple deforestation, oil investment naturally gives rise to carbon emissions via car exhaust.

Needless to say, Amazonian Indians are dead set against García's free-trade schemes with the United States. The spiritual leader of the Achuar Indians of northeastern Peru, who wears a bright red headdress made of toucan feathers and has red war paint streaked on his face, remembers how everything changed when the oil companies arrived. The animals ran away, the fish died, and crops started to wilt, he says. "The Peruvian state just wants to extract as much oil as they can from our land. They've made millions of dollars but we haven't seen it here," he adds.[32] The Achuar are unhappy about an oil company called Occidental that operated in a concession area of pristine rainforest known as block 1AB, located in the Corrientes River Basin. Over the course of thirty years the company discharged "formation waters," an untreated toxic by-product of the oil-drilling process, right into the rainforest. The Indians say that for every barrel of oil there were nine barrels of contaminated water produced as a by-product, which adds up to more than a whopping million barrels a day.[33]

Try to challenge the logic of oil investment and free trade with the United States, however, and you come up against some powerful forces. When the Indians claimed that García's land decrees were unconstitutional, the president was intransigent and refused to back down. How could he? Once he inked the free-trade deal with the United States, García had essentially locked Peru into an impossible and undemocratic arrangement. If the government suddenly had a change of heart and decided to protect the constitutional rights of the Indians, then García would have to confront powerful U.S. investors who would complain that free trade was being restrained.

Concerned that their lands would be privatized and subjected to deforestation, tens of thousands of Indians blocked Amazonian roads, rivers, and railways and occupied oil industry installations in an effort to pressure the president to repeal his free-trade decrees. García denounced the well-organized and peaceful protests, calling them terroristic and undemocratic. Characterizing the Indians as selfish, García said there was "a conspiracy afoot to try to keep us from making use of our natural wealth."[34] "These people aren't first-class citizens who can say—400,000 natives to 28 million Peruvians—'You don't have the right to be here.' No way. That is a huge error," García exclaimed.[35]

Digging in his heels, García escalated the situation by declaring a state of emergency, calling in the military and police, and charging protest leaders with crimes against the state. Descending on a group of indigenous people, including women and children, who were sleeping at a local roadside blockade, Peruvian special forces and helicopters dispersed the crowd with live ammunition and tear gas grenades. In the ensuing violence, twenty-five civilians and nine policemen died and more than one hundred people were wounded. The chief of police later claimed that the Indians were armed and fired first, but eyewitnesses said the indigenous peoples only had traditional spears and did not provoke the hostilities. Amazon Watch, a Washington, D.C.–based NGO, says the police came with orders to shoot. The organization interviewed witnesses who moreover claimed that the security forces later burned cadavers and threw dead bodies into the river.[36] In New York, three activists chained themselves to the door of Senator Chuck Schumer's office in solidarity with the protests in Peru. Schumer, they said, had blood on his hands. Before the U.S.-Peru Free-Trade agreement came up for a vote, activists had expressed concern to the senator that the agreement would spark violence in the Amazon, a warning that Schumer failed to heed.

Appalled at the tragic loss of life, Peru's congress voted 82 to 14 to revoke García's decrees and the president was forced to retreat. The president offered to negotiate with Indians, canceled the state of emergency, and even admitted mistakes in the handling of the protests. But while the legislative maneuvers temporarily defused tensions, some wonder whether root conflicts have been addressed, such as how to develop the Amazon and stop deforestation in the interest of halting climate change. In the wake of the negotiations, the government seems to be backtracking and gave the green light to an Anglo-French company to drill for oil in the Amazon. One environmentalist remarked, "Anyone who hoped that the dreadful violence of the past few weeks might have made Peru's government act with a bit more sensitivity toward the indigenous people of the Amazon will be really dismayed at this news . . . The government is trying to present a more friendly image in public, but as far as the oil companies are concerned, it looks like business as usual."[37]

As recent history has shown, affluent countries are hardly blameless when it comes to tropical deforestation in countries like Peru. From logging to illicit substances, consumers in the Global North are complicit in environmental damage that is adding to climate woes. Perhaps if rich nations succeed in passing a meaningful REDD initiative, then this will partially make up for the north's environmental sins. Now, however, the energy imbroglio adds another unwanted dimension to our environmental problems and complicates matters yet further. Though the Amazon is being exploited for natural resources like oil, it's also home to so-called "clean energy" economic development—projects that will also add to global warming.

Chapter Seven

THE "CLEAN ENERGY" TRAP

W hat will it take for rich countries to transfer clean technology to the Global South? That is the question many Brazilians are asking themselves as the authorities move forward with wasteful energy schemes in the Amazon. In Brazil, there has been a great move toward so-called "clean energy" initiatives, including hydropower and biofuels. Touted as alternatives to the fossil fuel paradigm, these technologies are hardly a panacea for our looming energy crisis and indeed will worsen our climate change difficulties. As we advance in crucial climate change negotiations, the Global North is going to have to do much more to invest in truly green technology such as wind, solar, and waves and provide vital expertise to such nations as Brazil.

If there's one thing that Brazilians can't stand, it's going without soccer, and President Lula is determined to avoid a repeat of the blackout or *apagão* that crippled the country in 2001. In that year authorities were forced to decree emergency measures, including a ban on power-hungry floodlights. A special government task force (nicknamed the "Blackout Ministry") called for the switching off of lighting on streets, beaches, and squares. In the midst of the energy crisis some Rio business leaders feared a crime wave and called for the army to be deployed in the event of power cuts. Meanwhile, panic-stricken citizens stocked up on candles, generators, and flashlights. When the rationing went into effect, cutbacks obliged schools and businesses to close and disrupted transportation, trade, and leisure. As street lighting in most major cities was cut 35 percent, police night shifts were increased and night games of soccer were prohibited.[1]

Determined to avoid the fate of his predecessor Fernando Henrique Cardoso, who suffered political fallout as a result of the blackout of 2001, Lula has said that developing hydro power in the Amazon is essential if the country wants to sustain more than 5 percent growth. The mere fact, however, that Brazil is afflicted by chronic energy problems does not mean that Lula must sacrifice the rainforest to hydro power and thereby intensify climate pressures. Indeed, critics charge that Lula's dam building is merely designed to satisfy big business, which gobbles up energy so as to export tropical commodities.[2] Indeed, soy tycoons like Blairo Maggi have supported hydro power expansion as a means of powering the agribusiness revolution in Brazil. Politically savvy and well connected, the soy barons have been successful in pushing their particular development vision in states like Mato Grosso.[3]

For years, ecology professor Dr. Philip Fearnside at Brazil's National Institute for Research in the Amazon has been conducting groundbreaking research on the environmental effects of hydroelectric dams. As I make my way to Fearnside's office in Manaus, I wonder how this Amazonian city gets its power. Hardly a honky-tonk jungle town like Iquitos, Manaus has a distinctly urban feel to it—there are highways, traffic jams, and large shopping malls. In the 1960s, Brazil's military government set up special tax incentives in Manaus and a large number of factories, particularly home appliance manufacturers, set up shop here. Once the city was declared a free-trade zone, Brazilians flocked to Manaus to stock up on affordable electronic goods. If you spend any length of time in frenetic downtown, you can't help but be struck by the sheer volume of goods that gets unloaded from trucks day and night and carted into local downtown stores. An old trading post, Manaus has grown from a city of a few 100,000 fifty years ago to a metropolis of almost 2 million today.

Amid the local bustle it's easy to forget that you're geographically isolated from the rest of the world by jungle. Servicing the energy needs of this industrious but remote business hub has been a challenge for successive Brazilian governments. In the 1970s the military authorities ordered the development of the Balbina hydroelectric dam, located some ninety miles to the south of Manaus. Designed to harness the power of the Uatuma River, an affluent area of the Amazon, Balbina represented a key pivot in the rush to develop Brazil's vast tropical rainforest. A full one-third of the Waimiri-Atroari tribe's population was displaced.[4] After the dam was filled, the water released through the turbines was virtually devoid of oxygen and this resulted in the killing of nearby fish.[5] A grandiose boondoggle, Balbina cost $700 million and experts reported that the expense was totally out of proportion to the energy yielded by the project: In the end, Balbina only produced 250 megawatts—enough to fill one half of Manaus's energy needs.[6] The city could have received electricity from other, more distant dams or oil and gas deposits instead of relying on Balbina.

Fearnside is concerned about all of these downsides, adding that dams also encourage road-building, which in turn leads to deforestation and land clearing by speculators and ranchers. Those displaced by hydroelectric projects move to new areas, where they clear additional forest.[7] In addition it's thought that dams alter the hydrological cycle in the Amazon, which in turn has an impact on precipitation in the Andes.[8] What really concerns Fearnside, however, is the climate change connection. The problem with dams, argues the scientist, is that they lead to emissions of methane, which are formed when vegetation decomposes at the bottom of reservoirs devoid of oxygen.[9] The methane is either released slowly as it bubbles up in the reservoir or rapidly when water passes through turbines.[10] According to Fearnside, Balbina flooded about 920 square miles of rainforest when it was completed and during the first three years of its existence the actual reservoir emitted 23 million tons of carbon dioxide and 140,000 tons of methane.[11] Fearnside has calculated that during this time Balbina's greenhouse gas output was four times that of a coal-fired plant producing the same amount of power.

News of the high methane emissions was particularly troubling as methane is twenty times more powerful a greenhouse gas than carbon.[12] Environmentalists say that methane gas produced by forests inundated by hydroelectric projects accounts for one-fifth of Brazil's greenhouse gas contribution to global warming.[13] How concerned should we be about dams and their effect on Earth's climate? According to researchers, the world's reservoirs release 20 percent of the total methane from all known sources connected to human activity, including livestock, fossil fuels, and landfills.[14] Environmentalists say that same methane released by dams, meanwhile, accounts for 4 percent of total global warming.[15] Researchers report that reservoirs contribute approximately 4 percent of all carbon dioxide emissions resulting from human activity.[16] Recently the issue of hydro power has been climbing up the political agenda of the world's leading scientists: In 2006 the Intergovernmental Panel on Climate Change (IPCC) included emissions from artificially flooded regions in its greenhouse gas inventory.[17]

In light of the negative environmental experience at Balbina one might think that the Brazilian authorities would take a harder look at hydroelectric power. Instead, the exact opposite seems to be occurring: Recently Brazilian authorities and business leaders have pushed hydro power development and in particular a controversial dam project called Belo Monte. Scientists have raised the alarm bell about the complex, which will cause an increase in greenhouse gas emissions due to the decomposition of organic matter within the stagnant water of the reservoir.[18] Belo Monte is located in Vitória do Xingu, in the state of Pará, and has been in the works since the 1980s, when it was originally referred to as Kararaô. Though the project was temporarily suspended in response to popular pressure, a hydroelectric company called Electronorte reinitiated work in the area in 2000. Presenting Belo Monte to society at large as a done deal, the company pursued an aggressive PR campaign financed with public funds.[19] In its advertising, Electronorte even went as far as to call Belo Monte a "blessing from God."[20]

Opposing these large-scale projects is enough to get you killed, as Ademir Alfeu Federicci tragically found out. An ecologist and land activist, Federicci, also known as Dema, was a leader of the Movement for the Development of the Trans-amazon and Xingu, a coalition of 113 indigenous, women's, religious, and environmental groups.[21] Scourge of the powerful establishment, the group worked to organize people living along the Trans-amazon Highway, the only road that runs east to west through the Amazon region. In the long term, the organization sought to promote sustainable development and roll back large hydroelectric projects like Belo Monte, which is expected to be completed by 2014 and would be the world's second largest dam.

Courageously, Dema denounced corrupt Xingu politicians and government figures linked to Electronorte, the hydroelectric company pushing Belo Monte.[22] One month before Dema's death, the thirty-six-year-old activist helped to draft a letter protesting the government's development plans for the region. Rather than pushing rash hydroelectric projects, the activist argued, the government ought to consider other economic alternatives such as eco-tourism and fishing.[23] Dema's call resonated among many who are still worried that Belo Monte will flood areas in and around the town of Altamira. Around Altamira, thousands of local residents live in homes built on stilts. The area floods during the rainy season but could be completely inundated once the dam is built. Some people have come to town fleeing *pistoleiros,* ranchers, and land grabbers in the countryside. They fear Belo Monte as they don't have anywhere else to go.[24]

A brave man, Dema charged that Belo Monte itself was rammed through with an air of "authoritarianism" characteristic of earlier Amazonian dam projects. According to the activist, the Brazilian police and intelligence services even spied on organizers and filmed public meetings organized against Belo Monte.[25] In a lawless area like Altamira, Dema took a big risk in organizing against the dam project. Land feuds and murders are common in the area, and some believe that the local police turn a blind eye to killings and crime. When federal agents finally arrived in Altamira, activists were relieved.[26] Indeed, Dema even helped federal police investigate the state governor, who had embezzled development money and channeled funds to ranching and other business interests.[27] Nevertheless, in general the federal police are grossly understaffed in the area and have few resources. What's more, they know their limits in a region where the majority of people own guns.[28] In August 2001 an armed man entered Dema's apartment and killed the activist.[29] The local police claimed that Dema was killed during a bungled burglary.[30] After carrying out some preliminary detective work the police picked up a local thief. After they got his confession, they threw the suspect in jail.[31] Dema's colleagues and federal police officers have said they believe someone hired the attacker to kill Dema.[32]

Facing repression and a hostile political milieu, anti-dam activists in the Amazon surely face difficult challenges. Yet, one prominent figure in the upper echelons of government had their backs: Marina Silva. From her perch at the Ministry of the Environment, the former rubber tapper watched with concern as the Lula government provided the green light for hydro power expansion throughout the Amazon. Specifically, the president sought to push through two dams on the Madeira River in the Western Amazon near Bolivia. Known as Santo Antônio and Jirau, the projects will jointly produce more than three times as much energy as the famous Hoover dam in the United States.[33] The Madeira River is the Amazon's longest tributary and one of the best-preserved jungle corridors on Earth. While the dams stand to provide up to 8 percent of Brazil's energy needs, they also threaten to undermine the region's fragile hydrology, worsen water quality, squeeze the aquatic food chain, encourage malaria, displace thousands of people, and contribute to chaotic growth in nearby urban centers.[34] But most importantly the projects could worsen climate change: Scientists say the dams will flood more than 200 square miles, thus releasing greenhouse gases from rotting vegetation.[35]

Hardly ingratiating herself amongst Lula's supporters in big business, Silva remarked that there was "no timetable for approving the Madeira project." Then IBAMA, an agency falling under Silva's purview, refused to give the environmental go ahead after citing concerns over sedimentation, flooding, and the potential for dredging up toxic mercury deposits, a byproduct of illegal gold mining.[36] Determined not to let IBAMA stand in the way of the Madeira project, Lula simply split the government agency into two, creating one entity dedicated to approving environmental projects and another focused on conservation.[37]

Having shunted aside the progressive forces in his government, Lula gave the green light to Madeira. In September 2008 construction began at Santo Antônio and by 2012 the government hopes to complete the Madeira projects, which could wind up costing up to $13 billion.[38] Where is all the money coming from for these hydroelectric boondoggles you ask? One chief culprit is the Brazilian National Development Bank, the financial arm of the Ministry of Development, Industry and Foreign Trade, which has been tied to the climate-unfriendly cattle industry.[39] In 2006, the bank pledged to finance up to 75 percent of the Madeira project.[40] And just in case you thought methane-producing dams were a strictly Brazilian affair, consider that the Inter-American Development Bank (IADB) is expected to also contribute to the Madeira project despite heavy lobbying from environmental and human rights groups that have been urging the bank to steer clear of the dam.[41]

With all of the social drawbacks associated with hydro power, not to mention the implications for climate change, why won't authorities consider meeting Brazil's future energy needs through alternative means? Recently environmentalists argued that the Lula government should upgrade existing energy systems and push through rapid development of wind, solar, and biomass technologies.[42] If Lula adopted such clean technologies Brazil could meet its electricity needs through 2020 and actually save $15 billion in the process.[43] Sounds like a proposal worth exploring, but predictably the electrical sector wasted no time in attacking environmentalists for being utopian and naive. To retrofit older dams and cut transmission losses was simply wishful thinking, the powerful lobbying group charged.[44] One expert says that hydroelectric projects die hard in Brazil. "It's like a Dracula movie," he says. "Every 20 years or so, it surges up out of the coffin. You have to drive the stake back through the thing and make it go away again."[45]

Politically and economically, the hydro power lobby is a juggernaut in Brazil. And yet, there may come a day when hydroelectric projects are simply not seen as environmentally sustainable anymore. Consider: The immediate cause of the 2001 energy crisis and blackout was a severe drought—the worst in more than sixty years. When the dry spell hit, water levels at hydroelectric plants fell to less than one-third of capacity.[46] In the long run hydro power may be caught in a vicious cycle of its own making: As large boondoggle projects such as Belo Monte proliferate, they may emit harmful greenhouse gases and thus contribute to climate change. This in turn could spur drought and more of an intense El Niño phenomenon. If global warming dries up parts of the Amazon, Belo Monte and other dams like it could wind up being white elephants as there won't be much water left to harness.[47]

In our zeal to make the transition away from fossil fuels, could we wind up worsening our already serious climate problem? That's the question now facing the planet, with the proliferation of so-called "biofuels" held out as a solution to our looming energy crisis and the prospect that we have reached—or may soon reach—peak oil. Supposedly, growing fuel-producing plants is more ecologically sustainable than oil and gasoline that pumps carbon dioxide into the atmosphere when burned. The idea is that plants can capture the energy of the sun and produce substances—such as sugars, starch, oil, and cellulose—that can be harvested and then converted into energy sources for us to use. In theory, plants take up carbon dioxide when they grow, and as a result burning fuel from them should not contribute to global warming.

Propelled by increased anxiety over skyrocketing oil prices and climate change, biofuels have become the advanced guard of the green-tech revolution and a fashionable way for corporations to demonstrate their commitment to alternative energy. Currently the United States and Brazil are the world's top producers of biofuels, accounting for 70 percent of global supply. In the United States, much of the gasoline that is sold is about 10 percent ethanol, a fuel produced from crops like corn. Brazil on the other hand, makes ethanol from sugar cane. Biofuels reduce dependence on imported oil while creating rural employment and enriching farmers.

Politicians like to push biofuels: It makes them look like they're being proactive on climate change but doesn't force them to call voters to account for their wasteful and consumerist lifestyles. But in practice biofuels may not be any greener than hydro power. Indeed, recent studies have shown that biofuels are actually accelerating global warming and imperiling the Earth in the name of rescuing it.[48] One example of this is our modern-day use of nitrogen-rich fertilizer in farming, which has increased sixfold over the last forty years. Unfortunately, not all the extra nitrogen winds up in crops: The fertilizer also yields nitrous oxide, known as laughing gas.[49]

After decades of research, Dutch scientist Paul Crutzen uncovered just how dangerous nitrous oxide is: It warms the planet at 300 times the rate that carbon dioxide does.[50] For Crutzen, fighting for political and environmental justice comes naturally. Motivated by his childhood experience in Nazi-controlled Holland, he has tackled a variety of environmental issues, from tropical cyclones to the devastating effects of nuclear warfare. Biofuels would not result in as devastating an environmental collapse as nuclear warfare, but that didn't stop Crutzen from doggedly pursuing the issue of nitrous oxide and its long-term implications with equal zeal. Most importantly from the point of view of climate change in the tropics, he released a study in 2007 finding that the release of nitrous oxide through farming can contribute toward global warming as much as or more than normal fossil fuels.[51] Crutzen's work challenged earlier standards used by the Intergovernmental Panel on Climate Change (IPCC). According to Crutzen, microbes convert much more of the nitrogen in fertilizer to nitrous oxide than was previously believed—3 to 5 percent. That's twice the widely embraced figure of 2 percent that the IPCC uses to calculate the climatic impact of fertilizers. As it turns out, growing many common biofuels released twice the amount of nitrous oxide into the atmosphere as previously thought.[52]

To get a sense of how biofuels have radically transformed the American countryside one need merely head to the state of Iowa. While the rest of the country frets about falling real estate prices and a worsening economy, parts of the rural United States have prospered in these hard times. Across northwest Iowa, corn farms cover tens of thousands of acres. In recent years the price of corn has shot up as a result of insatiable demand from local ethanol plants. That has benefited Galva, a sleepy agricultural town of less than 400 people that was transformed when tractor trailers began to spill golden kernels of corn into receiving pits at Quad County Corn Processors, the local biodiesel plant. The plant opened in 2002—since then it has converted millions of bushels of corn into ethanol with the aid of a generous federal tax credit that benefits those who produce and sell the fuel.[53] Farmland in the vicinity, meanwhile, shot up in value.[54]

The newfound prosperity shocked many. Up until recently corn prices had barely budged, averaging about $2 a bushel. The malaise prompted some farmers to forsake farming, leaving their native towns in decline. When corn prices went up, farmers earned more money, and local residents hoped that one day the agricultural sector would be free from government price supports. Travel about thirty miles northwest of Galva and you arrive in Marcus, another town riding the biofuels boom. Not only does Marcus have an ethanol plant but it's also constructing a biodiesel plant and a motel on the outskirts of town. The local mayor is thrilled that a truck stop has just opened up: It's a big deal in a town of just 1,100 people. For the first time in decades, young families have been moving back to rural Iowa as opposed to fleeing the area.[55] Recently, the ethanol industry has experienced an economic slowdown as a result of the recession, but if there's a lasting boom Iowa could indeed become the Texas of the ethanol industry.[56] As long as Washington mandates the continued use of biofuels, the future could be prosperous for these towns that have thrived on the production of renewable energy.[57]

But just how environmentally friendly is corn ethanol? It takes eight gallons of fossil fuel to grow, harvest, and convert crops to ethanol, so for every ten gallons of corn ethanol produced, there's really a net gain of just two gallons of renewable energy. There's another fact that farmers and their political allies conveniently choose to omit: Ethanol contains only about 70 percent of the energy of gasoline. Farmers counter that producing corn ethanol reduces the release of carbon dioxide. Furthermore, they argue, carbon dioxide that is emitted during the burning of ethanol is "canceled out" by carbon dioxide absorbed by the next crop of growing plants, which use it in photosynthesis. But think about all that carbon that's released when the prairie grasses and trees that once overran the land are turned over to corn production.[58] Indeed, in some cases the carbon released by corn cultivation will not be offset for at least a full fifty years by the gain attributed to biofuel generation.[59] It's a key point to bear in mind since Iowa farmers are already converting land set aside for environmental protection to grow crops.

Could we be concentrating too much on carbon while forgetting all about the 800-pound gorilla in the closet? There's another ecological downside to ethanol: nitrogen fertilizer.[60] The problem is that corn is "nitrogen leaky": Its shallow roots only take up nitrogen a few months out of the year and the rest is expelled into the atmosphere.[61] According to Crutzen, for corn ethanol the relative warming due to nitrous oxide could be .9 to 1.5 times larger than the relative cooling effect due to saved fossil carbon dioxide emissions.[62]

Try to challenge the ethanol industry these days and you run up against some heavy hitters. While many local farmers have benefited, the largest player has been agribusiness giant Archer Daniels Midland or ADM. With the help of federal largesse and subsidies, ADM has grown to a $44 billion-a-year operation and today is one of the world's chief buyers, sellers, and processors of grains and biofuel crops. The company produces ethanol at its cavernous Decatur, Illinois plant which stretches more than a mile. ADM has enormous strategic advantages in the region: The company operates a huge network of grain elevators around the Midwest, which permits the firm to purchase lots of corn at market lows. ADM also controls fleets of train cars, which are handy to bring grain to plants as well as ethanol to markets. For decades ADM was run as a virtual family fiefdom under the iron-fisted control of Dwayne Andreas, one of the country's most politically influential CEOs, known to presidents and prime ministers alike.

Ethanol producers have political support at the highest echelons of power in Washington, D.C. From farm-belt Republicans to Democrats, ethanol is being embraced as a way to reduce U.S. dependence on imported oil and generate economic opportunity in neglected rural regions. Republican Iowa senator Chuck Grassley has been a stalwart champion of ethanol on Capitol Hill. A soybean and corn farmer himself, Grassley created tax credits for ethanol and hopes that by 2025 Americans will derive 25 percent of their power from renewable sources.[63] The senator is passionate on the subject of biofuels and defends the industry against critics who charge that ethanol has led to higher food and fuel costs. In the spring of 2008, Grassley held a news conference and brandished a large box of Corn Flakes. Farmers in Iowa, he said, made less than ten cents from the box, which he bought for five dollars. "When a farmer gets so little out of a box of Corn Flakes, don't be blaming the farmer, and ethanol, for the high price of food," the senator remarked sternly.[64]

But even as he pledges his support to the ethanol industry and agribusiness, Grassley fails the environment. Republicans for Environmental Protection, a group that hopes to reform the GOP by fighting for the conservation ideals of earlier Republicans such as Teddy Roosevelt, has awarded Grassley pitifully low environmental scores based on his voting record in Congress.[65] Perhaps realizing that it was time to reverse his environmental reputation, Grassley threw his weight behind the development of an actual indoor rainforest in Iowa.[66] Known as Earthpark, the project was to be constructed on an industrial site located in the town of Coralville, a thriving community near the University of Iowa and "The World's Largest Truck Stop" along Iowa 80. The center is intended as a national tourist destination and will feature recreations of natural ecosystems, including a 4.5-acre rainforest.[67] The tropical main attraction will be located under a three-story, translucent dome.[68] Visitors will be able to actually walk through the rainforest canopy on catwalks while surrounded by such captivating creatures as poison arrow frogs, capuchin monkeys, and sloths.[69] In addition, organizers plan to feature plants such as the açaí palm, itauba, and rubber tree, which will help to form a more complex rainforest canopy. Tourists can view blue and yellow macaws while wondering at the exhibit's aquatic life, including electric eels, freshwater sting rays, and several species of piranha.[70]

Sponsors of the project hope that Earthpark's "iconic architecture" will lure thousands of motorists from nearby I–80 and become as recognizable a symbol as the Gateway Arch in St. Louis or the Seattle Space

Needle.[71] Philippe Cousteau, an Earthpark board member and the young grandson of the famous filmmaker and natural explorer Jacques Cousteau, is thrilled that the center will showcase solutions for putting a break on global climate change.[72] Sounds educational enough, but Earthpark has been held up for years due to lack of funding. Though Grassley helped to funnel a $50 million federal appropriation to the project through the Department of Energy in 2003, boosters have had trouble raising the remaining $90 million.[73] Earthpark has become the butt of local jokes and the "Iowa Pork Fest" has made the network news.[74]

Sponsoring Earthpark sounds all fine and good, but if Grassley is concerned about what is happening to the actual rainforest then he might want to reexamine his support for increased corn production and ethanol. The Brazilian *cerrado* and Amazon rainforest lie thousands of miles away from small Iowa towns such as Galva and Marcus, but what happens in the American Midwest will have drastic consequences for Brazil's future. Call it the law of unintended consequences: When the price of corn shoots up, more Iowa farmers switch from soy cultivation to corn. That in turn leads to a rise in the global price of soybeans and encourages more farmers at the world-wide level to plant soy.[75] In particular, increased demand for ethanol in the United States has likely contributed to Brazilian farmers deforesting the Amazon to plant more of the world's soybeans.[76] The *cerrado* has not been immune, either, from such international pressures. Indeed, when soy displaces ranching, ranching moves into unspoiled areas of the *cerrado*.[77] As mechanized farming penetrates the *cerrado* and encourages greater carbon emissions, land prices have gone through the roof, just like in Iowa.[78]

Not that agribusiness in Brazil needs to be pushed much by outside sources and international commodity prices to develop: These days the South American country has its own vast biofuel industry that is spreading throughout sensitive areas and worsening our carbon conundrum. When you fill up your car with a gasoline and ethanol blend most likely you are burning ethyl alcohol produced from U.S. corn. A few years from now, however, your commute may be powered by ethanol made from sugar cane cultivated in the Brazilian *cerrado*. An economic powerhouse dynamically bursting forth onto the world stage, Brazil is the Earth's largest producer of sugar cane ethanol. President Lula has jumped on the ethanol bandwagon, repeatedly remarking that his country's fortunes depend on a future in which "we plant and harvest fuel."[79] In São Paulo, a sprawling city of eighteen million, motorists can fill up their tanks with either gasoline or ethanol, known in Brazil as *alcool*. Most opt for ethanol, not surprising given that this biofuel costs half the price of gasoline.

Though Brazil's exploration of ethanol production goes back to the 1920s, it wasn't until the 1970s that the industry started to gain any traction. Buffeted by the oil price shock of 1973, the country's military dictators grew concerned about Brazil's reliance on foreign imports of fossil fuels. Their solution: Pour government subsidies into the sugar industry and mandate ethanol distribution at the pumps. Today one can buy so-called "flex-fuel" cars (known as *carros flex* in Portuguese) that run on gasoline, ethanol, or any combination of both. Because ethanol is cheap, the public has bought up the cars, and currently a full 90 percent of new vehicles sold can run on *alcool*.[80]

Ethanol's not just keeping cars on the road; it's also keeping a small but growing number of aircraft aloft. Neiva, a subsidiary of aircraft manufacturer Embraer, has turned out some 200 single-engine, single-seat planes that run on ethanol. While biofuel technology is only being utilized for propeller planes right now, Brazil has 14,000 planes, the second-biggest fleet of light aircraft in the world after the United States, and approximately 12,000 of these could be adapted to use ethanol.[81] There's no escaping it: Brazil has become deeply committed to ethanol, and economists expect that "alcool fever" could attract some $100 billion in investment. The industry hopes that this will lead to the construction of one hundred new distilleries by 2012, by which time domestic ethanol production will have doubled.[82]

On the surface of it, Brazil's ethanol boom has proven to be both economically and environmentally desirable. The sugar cane-based fuel is 30 percent less expensive to produce than the corn-based variety coming from Iowa.[83] Sugar ethanol is also efficient: One acre of sugar cane can yield more than twice as much ethanol as an acre of corn.[84] Compared to corn, sugar cane also looks pretty environmentally friendly: Brazilian farmers use less fossil fuel to convert sugar cane to alcohol than Iowa's farmers do to produce corn ethanol. Even better, plant waste from sugar cane can be used to produce heat and electricity right in the distillery itself.[85]

As a result of its turn toward ethanol, Brazil avoided emitting 600 million tons of carbon between 1974 and 2004.[86] So what's with environmentalists who complain about ethanol—won't they ever be satisfied? While sugar cane ethanol is certainly less ecologically destructive than some other biofuels, the industry's boosters have overlooked one key fact: You've got to plant sugar cane somewhere. One couldn't pick a worse place to harvest cane than Brazil's Atlantic rainforest. There, sugar cane crops have led to deforestation and, paradoxically, more carbon emissions.

It's difficult to imagine that a serene and pastoral landscape lies just beyond São Paulo. Take a bus through the city and the miles and miles of grey industrial factories stretch on forever. But nearby is the Atlantic rainforest, also known as the Mata Atlântica. When the first Portuguese explorers stepped ashore in 1500 A.D., the forest may have covered more than 500,000 square miles, or approximately one-fifth the size of the current Amazon jungle lying 500 miles to the northwest.[87] To put it in perspective, that's an area about twice the size of the state of Texas. Located in the Brazilian south and southeast, the Atlantic rainforest ranged all the way up to the Northeast in a long coastal strip.[88] In some areas the forest even extended a full 300 miles inland or more and encompassed a broad spectrum of habitats, including coastal mangrove thickets and mountain massifs 3,000 feet high, covered in broad-leaved evergreens and conifers.

In a bad omen, one of the first things the Portuguese explorers did was to chop down a tree. They then built a cross out of it and celebrated Mass, claiming the land and rainforest for God and king. In short order the Portuguese went to work, cutting down trees and releasing the carbon stored in the rainforest.[89] In 1525, the Portuguese began to grow sugar cane and introduced the crop to the Atlantic rainforest.[90] Then, the colonists shipped six million African slaves to Brazil to do the cutting.[91]

Over the next 500 years the Atlantic rainforest bore the brunt of Brazil's economic development. The country's eastern seaboard has long been the main population and industry center—70 percent of Brazil's people live there and the area includes huge cities like São Paulo and Rio de Janeiro.[92] Over time, Brazil lost about 93 percent of the Atlantic rainforest and today only tiny remnants of the ecosystem remain.[93]

Today, Rio and São Paulo are congested mega-cities, yet try as it might Brazil cannot escape its colonial sugar legacy. Just outside the urban center ethanol producers have set up shop in the Atlantic rainforest, and last year the government fined two dozen of these firms for illegally clearing the land.[94] After the authorities clamped down the companies were obliged to restore 143,000 acres of rainforest.[95]

Whatever the environmental advantages of ethanol, this thriving business now threatens our Earth's climate balance by its destruction of the Atlantic rainforest. It is ironic that a supposedly green industry could wind up imperiling such a valuable habitat. Though it's a fraction of the size of the Amazon, the Atlantic rainforest contains a similar range of biological diversity. Consider: The area has about 2,200 species of birds, mammals, reptiles, and amphibians. There are almost 200 bird species and 21 species of primates in the Atlantic rainforest that are not found anywhere else in the world. Furthermore, there are approximately 20,000 species of plants, representing 8 percent of the world's total, and new species of flora and fauna continue to be discovered.[96] Among biological hotspots—environmentally threatened regions with a high number of species encountered nowhere else in the world—the Atlantic rainforest ranks as one of the top five.[97]

All of those trees that we depend on to store harmful greenhouse gases have a diverse biological portfolio. Botanists have found that the Atlantic rainforest has the highest known tree diversity in the world. Indeed, a team working with the New York Botanical Garden discovered a whopping 450 different species just within a two-and-a-half acre plot of forest. The discovery upended a previous density record of 300 species within a similar area along a flank of the Andes Mountains descending into the Amazon rainforest. To provide a bit of perspective on the matter: Only ten tree species typically inhabit a two-and-a-half acre plot in New England.[98] The dazzling tree diversity is the result of multiple canopies that support a rich mixture of vegetation, including ferns, mosses, and epiphytes, including lianas, orchids, and bromeliads.[99]

Those diverse trees in the Atlantic rainforest that we're relying on to regulate world climate also provide important habitats for local animals. Consider for a moment the case of the charismatic golden lion tamarin, an orange-yellow monkey and endangered resident of the area.[100] When it comes to monkeying around, this animal is second to none: Few species possess its impressive skill or agility. At night, the primates sleep in tree holes, which are important for heat conservation and protection from nocturnal predators such as hawks and other raptors, cats, and fearsome snakes. Retiring at dusk, the tamarin sleeps in the tree until sunrise. This primate also depends on trees for food: The tamarin pries up bark looking for insects and the like. Using their long, slender fingers and hands, the tamarin probes these hard to get to areas—it's a technique known as "micromanipulation."[101] But as the trees of the Atlantic rainforest come down the tamarin loses its home.

Yet another endangered species, the maned three-toed sloth, also depends on the Atlantic rainforest's trees for its livelihood. The rarest of sloth species, this creature has a low metabolism that enables it to survive on an energy-poor diet of tree leaves.[102] The maned sloth travels by hooking its long claws over branches, allowing it to easily maneuver among the trees.[103] During the day, the creature rests within tangled vines on tree tops, only emerging late at night.[104] The sloth's cream and tan coat is frequently tinged with blue green algae, which thrives in the grooves of the creature's hair. It's an ingenious system of camouflage that allows the sloth to hide in trees from predators such as jaguar and human hunters.[105] While these two are certainly major threats to the sloth, the animal may face an even greater danger in the form of land clearing for sugar cane.

Brazilians love powering their cars with cheap ethanol. The question is where the country should plant its sugar cane. The environmental destruction unleashed by ethanol in the Atlantic rainforest is troubling enough, but what if sugar cane were to lead to more deforestation in other sensitive areas? Today the Brazilian sugar cane industry is centered in the state of São Paulo—drive just an hour out of the city and you can see sugar cane fields stretching for hundreds of miles.[106] Palmares Paulista is a rural agricultural town 230 miles from São Paulo. Behind rusty gates lies a squalid red-brick tenement building. Inside, weary migrant workers breathe the stale air and try to prepare themselves as best they can for the long day ahead. The *cortadores de cana*, or sugar cane workers, are crammed into tiny cubicles filled with rickety bunk beds and unpacked bags. They hail from the poverty-stricken, drought-plagued northeast and earn paltry wages.

Like migrant workers in the Amazon who are mercilessly shanghaied to clear the rainforest, Brazilian *cortadores* work in a shadowy world of human rights abuses. Unscrupulous landlords, sometimes ex-cutters themselves, take advantage of the newcomers by renting out seedy flop houses at outrageously inflated prices.[107] Middlemen bus the workers to cane farms and charge stiff fees for transportation, food, and supplies. As a result, the *cortadores* fall into a miserable debt spiral and are forced into a form of slavery.[108] On the plantations, cutters are not allowed to eat anything save corn meal and water. In truth, the wages are insufficient to buy anything else.

At work the cutters employ a machete, or *facão*. The blade is wider than a hand, and the *cortadores* sharpen it seven or eight times a day. The hook at the end of the blade can make serious wounds, and, laboring in the blisteringly hot fields, workers frequently suffer cuts from the sharp cane itself. When one cuts the cane, one strikes twice with the *facão*, once to separate the cane from the root and then to remove the remaining leaves from the stalk. Then, the worker twists the stalk with his free hand. To be a good cutter, you've got to carry out fast and fluid strokes that require considerable strength. At the end of the day, after waving their machetes 3,000 to 4,000 times, the men are frequently so tired that they can't even speak.[109]

Because their pay is based on the weight of their cuttings, the workers must put in twelve-hour shifts or more.[110] It's very difficult to complain about working conditions—*cortadores* work under the iron rule of the *feitor*, or foreman. In the fields, the *feitor* determines the workers' salary, who is hired, and who gets fired. Step out of line on the job and you could run afoul of the sugar barons' militias, known as *capangas*. Officially the militias are security guards, but in reality they operate as a kind of paramilitary force. Feared by the workers, the *capangas* intimidate laborers and drive off small farmers with bulldozers. Crisscrossing the plantations in jeeps and on dirt bikes, the *capangas* carry radios and weapons. Occasionally, government investigators manage to liberate cane cutters on the plantations, but in a vast country like Brazil these officials are scarce.

Brazil's one million cane cutters last an average of twelve years on the plantation—then they are casually replaced. In the fields you've got to be careful to drink enough water, and many who fail to keep hydrated will actually collapse and die on the spot. In cemeteries near sugar plantations, the *capangas* and *feitores* have nice graves. The bodies of cane cutters, however, are usually only buried for two years. After that, overseers dig up the remains and cart them to the garbage dump, where they are burned. The overseer burns the bodies so as to avoid payment of a $7 annual fee for each gravesite—it's too costly for the widows of the cane cutters to pay.[111]

Despite the hellish conditions for the workers, ethanol has been able to sell itself to the public on its ability to reduce carbon emissions. Again, however, there are other greenhouse gases to consider besides carbon dioxide. The Brazilian ethanol industry uses more than 240,000 tons of nitrogen fertilizer per year at a cost of about $150 million.[112] At a public senate hearing in Brasilia called to discuss climate change, experts expressed concern that nitrogen fertilizers used in conjunction with sugar cane production yielded nitrous oxide.[113] What's more, when you cut cane by hand you've got to set controlled fires in the fields to smoke out razor-sharp leaves, nasty snakes, and tarantulas.[114] In the middle of the night, plantations look like a war zone as burning fields light up the sky and the wind blows billowing smoke clouds far and wide.[115] Not only do the burnings pollute the air with soot, causing a number of illnesses, but they also release methane, a potent greenhouse gas, and nitrous oxide.[116]

Public officials declare that ethanol will not lead to deforestation in the Amazon or exacerbate climate change. They say that the particular soils and rainy weather characteristic of the rainforest are not suitable for the growth of sugar cane.[117] Agriculture minister Reinhold Stephanes has been quoted as saying that "Cane does not exist in Amazonia." In a withering blow to Stephanes's credibility, however, authorities recently raided a sugar cane plantation in the state of Pará where 1,000 workers were laboring under appalling debt slavery conditions.[118] In all, environmentalists claim, hundreds of thousands of acres of sugar cane have been planted in the Amazon.[119]

Even if there are only a few cane plantations operating in the Amazon, ethanol may exert an indirect impact on the rainforest through a phenomenon known as "agricultural displacement."[120] Though the state of São Paulo is located far from the Amazon rainforest, the sugar cane there can drive other crops toward the agricultural frontier. In the state of São Paulo, sugar cane has been planted on former pastureland and this has pushed cattle into Mato Grosso.[121] Hundreds of thousands of cattle are moving into the Amazon every year as a result of displacement by ethanol in the state of São Paulo alone, say environmentalists. This migration is becoming all the more likely since one can purchase 800 hectares of land in the Amazon for the price of just one hectare in São Paulo.[122] Additionally, some soy plantations in the center of the country have been turned over to ethanol production, prompting concern among environmentalists that this will lead soy producers to move into the Amazon.[123] And local observers say that sugar cane plantations are already pushing soy farmers and ranchers into the rainforest.[124]

Despite its many social and environmental costs, large financial institutions in the Global North have offered their support to the biofuel industry. There is big money to be made in this new industry: Worldwide investments in biofuels are expected to exceed $100 billion by 2010. Take, for example, the case of the IADB, which has provided funding to the Brazilian ethanol maker Nacional de Açúcar & Álcool to the tune of $153 million.[125] The IADB venture also features the Carlyle Group, a multi-billion-dollar equity investment firm with ties to the energy and defense sectors.[126] Sometimes referred to as the "Ex-Presidents Club," the Carlyle roster has included top figures from the first Bush administration and former President George Bush Sr. himself worked as a senior adviser to the company.[127] Moreover, George W. Bush briefly served on the board of Caterair, an airline catering company bought by Carlyle in the 1990s.[128] Then there's HSBC, a large London-based bank with global reach, which provided loans to Pará Pastoril e Agrícola, which grows cane for ethanol on a 24,700-acre plantation in Parás Amazon. There, workers were found in horrendous debt-slavery conditions.[129] In addition, well-known financiers and companies are getting into the ethanol market, from Richard Branson and George Soros to GE and BP to Ford and Shell to Cargill. In a sign of the times BP has conducted a PR blitz touting the virtues of ethanol. The company now says its initials stand for "Beyond Petroleum," and in full-page newspaper ads announces that it is time to produce more energy in the Midwest as opposed to the Mideast.[130]

Once the power elite in Brazil and the United States starts down a particular energy path it becomes very difficult to shift course. Despite the fact that George Bush had longstanding links to Texas oil, the former president did much to solidify South American ethanol ties. Why? Geopolitics may provide part of the answer. In 2005, Venezuelan president Hugo Chávez was at the height of his political popularity. An outspoken critic of U.S. foreign policy and corporate-friendly "free trade," Chávez used his country's vast petroleum reserves to undermine Washington's influence throughout South America. At the Summit of the Americas held in Mar del Plata, Argentina, before a crowd of 25,000, Chávez famously baptized the site as the "graveyard" of Bush's Free Trade Area of the Americas (FTAA). The summit ended in a fiasco: Bush left Argentina empty handed without any trade deal.

Determined to deflect Chávez's rising petro power, Bush headed to Brazil. There he received a much warmer reception: President Lula invited him to his own country home, known as the Crooked Creek Ranch.[131] The Brazilian authorities were keen on expanding the country's domestic ethanol base and promoting massive world-wide biofuel exports. Ambitiously, the South American giant sought to produce enough ethanol to substitute 10 percent of world gasoline demand in less than twenty years.[132] For the Brazilian president, trading conditions with his northern neighbor was vitally important: Though the United States imported billions of liters of ethanol every year from Brazil, Washington had imposed a stiff tariff on the biofuel.

Eating at a relaxed barbecue the two men developed a personal rapport. Prior to the meeting, Lula had instructed Minister of Agriculture Roberto Rodrigues to make the case for biofuels. Rodrigues, himself a soy farmer, spoke for a full hour during the barbecue. "How is it that humanity built a civilization upon fossil fuels, a finite substance that is poorly distributed around the world?" he asked. "It makes no sense when we have a renewable liquid that can be produced by almost any country." Reportedly, Rodrigues piqued the Americans' interest and Bush and Secretary of State Condoleezza Rice peppered the minister with questions. Bush officials later remarked that the president returned to Washington "all charged up" about Brazilian biofuels.[133]

Perhaps personal ties also help to explain Bush's enthusiasm for biofuels. Indeed, the president was getting an earful from brother Jeb, the governor of Florida. In 2006 Jeb paid a visit to the historic Biltmore Hotel in Coral Gables to meet with Roberto Rodrigues and Luis Alberto Moreno, the president of the IADB. At the time, Brazil was Florida's top trading partner to the tune of more than $10 billion per year, and Jeb was intent on deepening trade ties. In an effort to promote the use of cane-based ethanol throughout the Americas, Jeb called for the creation of the Inter-American Ethanol Commission.[134] During the proceedings a reporter asked the Florida governor if he regretted not launching his ethanol initiative earlier in his term. "It's human nature to do the same thing over and over again," Bush said, referring to our addiction to oil. "It's the way we are." On the other hand, Bush declared, it was now up to the United States "to do the uncommon thing, and look over the hori-zon."[135] By including members of the Florida FTAA on the commission Jeb was also able to rally his free trade supporters—the kinds of people that Hugo Chávez loved to hate.[136] Following up on his meeting, the Florida governor sent his brother George a nine-page report urging him to cultivate ties to friendly countries like Brazil. Jeb urged the president to more than double American consumption of ethanol by 2015 so as to reduce dependence on Venezuela, a country that supplied about 12 per-cent of U.S. petroleum needs.[137]

As a result of Rodrigues's lobbying efforts, Brazil seemed poised to take off as a massive regional energy hub. Agribusiness giants Archer Daniels Midland and Cargill—the same companies engaged in ethanol production in the American Midwest—now rushed to set up shop in Brazil to take advantage of the sugar cane bonanza.[138] Talk to Brazilian officials and they'll tell you that none of this exerts any pressure on pris-tine areas or exacerbates climate change. They claim that at least 90 mil-lion hectares are available for cultivation outside of the rainforest, land that can be used for a huge increase in ethanol production.[139] In practice, however, much of this agribusiness expansion is set to take place within the *cerrado*. Indeed, though approximately 60 percent of the country's cane is grown in the state of São Paulo, production has already been ramped up in the *cerrado* in Mato Grosso.

The *cerrado* is hardly an unused wasteland—it's a rich ecosystem in its own right. Or, at least it was until the advent of agriculture and ranch-ing. Today less than 50 percent of *cerrado* vegetation remains and defor-estation is occurring at about 1.5 percent per year.[140] One former legislator who helped to enact a law that mandated a 23 percent mix of ethanol be added to all petroleum supplies in Brazil now regrets sup-porting the ethanol cause in Congress. "Some cane plantations are the size of European states," he said, adding that "these vast monocultures have replaced important ecosystems."[141] If one takes the carbon dioxide released during land conversion into account, it would take sugar cane ethanol produced from previously wooded *cerrado* land a full seventeen years to repay its climate debt. But at that rate, growing sugar cane in the *cerrado* would never repay its climate debt since cultivating the crop frequently depletes the soil in less than seventeen years.[142] Not to worry, say government officials: Sugar cane will only be grown on grazing land that doesn't have natural vegetation.[143]

How will the government ensure that the more pristine *cerrado* doesn't get converted to sugar cane plantations? There are forestry laws on the books limiting deforestation in the *cerrado*, yet enforcement is poor. With everyone paying attention to the rainforest, the *cerrado* has been left out of the picture. That's allowed some high-profile financiers to set up shop in the area without anyone batting an eyelash. For example, AOL founder Steve Case and former World Bank president James Wolfensohn have invested in Brenco, the Brazilian renewable energy company, which seeks to produce 3.8 billion liters of ethanol in the *cerrado* by 2015.[144] As environmental concerns mount, the Europeans have become increasingly skittish on the question of biofuels and may even put the brakes on Brazil's ambitious energy expansion. Indeed, the European Union has been pushing for biofuel restrictions and is reconsidering its biofuel mandates. Stavros Dimas, the EU environment commissioner, has sought guarantees from Brazil that the ethanol it exports to the European Union isn't damaging the Amazon.[145]

There's been a fierce back and forth between European and Brazilian officials on the question of biofuels. The top scientist at the U.K. Department for the Environment recently warned that mandating more biofuel use as proposed by the European Union would be "insane," as this would lead to an increase in greenhouse gases.[146] Sweden, the only European country that already imports Brazilian ethanol for its public transportation system, used to think biofuels were heaven but now believes they are hell.[147] After allegations that some Brazilian sugar cutters were paid paltry wages, were underage, and even perished at a young age from exhaustion, Swedish motorists threatened to cease their use of this supposedly green fuel.[148] To make matters worse for the burgeoning Brazilian ethanol industry, the United Nations has added its voice to the chorus of critics. Achim Steiner, head of the body's environment program, declared that growing international demand for ethanol would threaten the Amazon if safeguards were not put in place.[149]

Shooting back at critics across the Atlantic, President Lula said that biofuels were actually an effective weapon in the struggle against global warming.[150] Lula chided Europe further, claiming that the developed world was simply jealous of Brazil's emergence as a major agricultural powerhouse. "Just when Brazil appears on the world stage not as a bit part actor but as the lead in a play about agricultural production . . . people start to get uncomfortable, very uncomfortable," he said.[151] Furthermore, the Brazilian politician remarked, European competitors were using the environment as a red herring to stall Brazil's biofuel industry. "We have adversaries that will make up any kind of slander against the quality of ethanol," Lula declared.[152]

George Soros is one prominent financier hoping that all the controversy swirling around cane ethanol will blow over. It is ironic that Soros is in the same camp with the Bush family and the Inter-American Ethanol Commission when it comes to biofuels. In the lead up to the U.S. presidential election in 2004, the financier launched a month-long, twelve-city tour against President Bush and the war in Iraq and donated more than $15 million to liberal activist groups with the express idea of ousting George W. Bush from the White House. An environmentalist and critic of unbridled capitalism, Soros has donated billions of dollars to promoting human rights and the rule of law around the world.[153] Yet, when it comes to promoting the biofuel boom in Brazil, Soros has much in common with Jeb and George Bush. He invested $900 million in Adeco Agropecuaria Brasil, a company that seeks to produce biofuel from corn and sugar cane grown in the *cerrado*.[154] Speaking at an ethanol summit in São Paulo, Soros said that he would continue to invest in Brazilian ethanol and added that the United States would have to bring down its tariff wall or ethanol would have a tougher time getting off the ground as a viable alternative fuel.[155]

Some progressive forces in Brazil would just as soon see Soros fail. In 2007, when President Bush visited the city of São Paulo to promote biofuel cooperation between Brazil and the United States and open talks about a kind of two-nation "OPEC of Ethanol," 10,000 people spilled into the financial center in opposition to the visit.[156] While city-dwellers banged drums and waved red flags in the broad avenues of their city, landless peasants working with the Landless Rural Workers' Movement (MST) carried out non-violent occupations of twelve sugar plantations and a Cargill-owned ethanol distillery in the state of São Paulo. Brazil ought to make ethanol, organizers said, but should not encourage large-scale industrialized production for export to the Global North.[157]

As protests raged around São Paulo, Bush and Lula coolly discussed ethanol's future in the Americas. Though the United States would not rescind its $0.54-per-gallon tariff on imported ethanol, the two countries signed an important agreement to develop alternative fuel sources by sharing common experiences and pooling technology. The IADB would again be involved, this time as a partner in Brazilian-U.S. feasibility and technical studies.[158]

One MST activist was disappointed. What was needed, he remarked, was a different agricultural model focused on the needs of small farmers and food production.[159] "Bush came to Brazil as a messenger boy for the multinational companies, the agribusiness companies, the oil companies and the automobile companies that want to control bio-fuels," the activist remarked.[160]

Environmentalists like Greenpeace, who occupied a large stone monument with a banner that read "Bush and Lula—Ethanol is Not Enough. Stop Global Warming," were also disappointed. For Bush to place such an aggressive focus on ethanol without mandating a cap on emissions or an increase in fuel efficiency for U.S. vehicles was simply a smoke screen, demonstrators charged.[161] Surveying the political landscape, MST activists sounded a note of caution: Through its relentless support of agribusiness and ethanol the Lula government would "provoke a popular reaction sooner rather than later."[162]

From the IADB to the Bush family to the Carlyle Group to George Soros to HSBC, there's plenty of blame to go around for the clean energy charade. Perhaps, if we successfully managed to capture methane from large dams within reservoirs and used it to generate electricity in power plants, as some scientists have proposed, then we could lower the negative climate impact of hydro power in the Amazon.[163] For the time being, however, neither hydro power nor biofuels look like they are going to solve our global warming problem in the tropics. If it wants to avert yet further climate change the Global North needs to get serious about the transfer of truly green technologies. Specifically, it would be much more promising to concentrate on other forms of clean energy such as wind, solar, and waves. After years of dragging its feet the United States is finally taking steps to promote such technologies. Indeed, the United States is now the biggest generator of wind energy world-wide, and the new administration in Washington has just finalized a framework for renewable energy production offshore.[164] Though the White House has proposed cutting research into tidal power, it has sought more funding to study solar energy.[165] If the Global North would do more to develop alternative clean energies, investing in a massive "Manhattan Project" for conservation, then new technologies could be shared with tropical nations such as Brazil that have been sacrificing the rainforest for short-term gain.[166]

For years, poor countries have been waiting for exactly this type of commitment from their northern counterparts.[167] In late 2008, at an international conference in Beijing, the Chinese called for technology transfer to be speeded up. During the meeting, the Chinese prime minister expressed frustration at the lack of an international mechanism that would ensure timely access to such clean technologies in the future.[168] China, along with other poor countries in the so-called "Group of 77," wants the United Nations to establish an executive body of technology that would be governed by many of these same nations. Such an executive body would determine technology-related financial requirements and ensure that privately owned technologies are available on an affordable basis. The United States, Japan, and the European Union would contribute to the fund. Needless to say such a technology transfer would come at a price tag—more than $140 billion a year from the United States alone.

In light of the historic responsibility of the Global North for global warming it's entirely reasonable for poor countries to expect a large financial commitment by rich countries, yet any mention of such a scheme elicits howls of protest from U.S. conservatives who oppose increasing foreign aid.[169] Indeed, throughout the Bush years Washington delayed on the question of technology transfers. In 2005, the United States, along with several other nations, agreed to a climate change pact advocating new technologies rather than emissions reductions. The deal, however, amounted to little more than a voluntary trade agreement in energy technologies and a repackaging of existing bilateral and multilateral technology transfer efforts that the United States had been pushing for years.[170] In the twilight of the Bush administration, however, the United States agreed to set up a multi-billion-dollar fund to help poor countries acquire clean power technologies. The fund was expected to draw financing from other affluent nations responsible for global warming.[171]

It's imperative that any deal be based on renewable energy—that is to say power drawn from wind, sun, tides, or heat from below the Earth's surface. Unfortunately the so-called "climate investment funds" had no clear definition of clean technology and left open loopholes through which environmentally dubious energy proposals, such as "clean coal" power plants, could be financed. Even worse, the fund would be run by the World Bank, an institution that spent much of its sixty-plus years promoting projects that caused pollution.[172] Green groups have called for public and private funds to be set aside in an independent global body and financed to the tune of at least $140 billion a year.[173]

Perhaps Obama will do more to address these sensitive issues, though up to now the signs have not been very encouraging. During recent climate negotiations in Bonn, Germany, the Obama administration showed little desire to assume leadership in the fight against global warming, and the Global North failed to agree to substantial transfers of money and technology for poor countries.[174] The lack of action prompted an outcry from China, which accused affluent countries of reneging on sharing technology. India, which has proposed creating global "innovation centers" to work on green technology, was also critical of the Global North.

All of this comes as a bad sign and suggests that access to green technology could become a growing stumbling block in world-wide efforts to fight climate change. Indeed, U.S. lawmakers bristle at the notion of China "stealing" U.S. technological knowledge. Just recently, in fact, the U.S. House of Representatives voted unanimously to ensure that climate negotiations in Copenhagen would not "weaken" U.S. intellectual property rights on wind, solar, and other green technologies.[175] Obama himself has not been much better than legislators on Capitol Hill—recently he stated that his administration would provide technological transfers but would not elaborate on the exact particulars of the financing.[176]

Who should transfer what, to whom, and at what price?[177] These are vital questions that must be addressed soon. Nations of the Global South, such as Peru and Brazil, are hoping that the rich countries of the world will offer them crucial green technology at climate negotiations in Copenhagen so they can avoid the worst effects of climate change.

Epilogue

FOOD AND FARMERS

IN AN ERA OF

CLIMATIC UNCERTAINTY

For many in the Global South, climate change will lead to more erratic and expensive food supplies. Indeed, chronic hunger may be the defining human tragedy of the twenty-first century as climate change causes growing seasons to shift, crops to fail, and storms and droughts to destroy fields. As the Earth's climate becomes more and more unpredictable, a crucial question for humanity will be: How do we secure crucial food supplies? While Canada and some other temperate-zone countries could benefit from global warming and actually grow more food—partly because rising temperatures will lengthen the growing season at higher latitudes—poor countries stand to lose out since extreme heat disrupts crop development. Already in South America farmers are observing that the regular seasons have been thrown off while the cycles of drought and flooding have made life more unpredictable.

What is more, as climate change intensifies it is spurring tremendous social conflict, particularly in the Brazilian northeast. For centuries, droughts have plagued these impoverished, semi-arid backlands, killing thousands at a time. To this day many northeasterners flee the parched area and wind up in the shantytowns of São Paulo and Rio de Janeiro. The northeast has also long been plagued by deep-rooted social injustice. Indeed, sugar mill owners disliked talk of land reform and any talk about this issue could get you killed. Historically the northeast has had the highest illiteracy rate in Brazil, and infant mortality and life expectancy rates compared unfavorably to the rest of the country. Chronic hunger has been a widespread occurrence with residents suffering from a deplorably low daily caloric intake. In the northeast, the word "pygmatization" has been used to describe the process in which people's growth was stunted as a result of a poor nutritional diet. During droughts infants died of dehydration and thirst-driven residents walked miles and miles in search of water. One- and two-year-old children could not sit up unaided and were incapable of speech. Their skin was stretched so tightly over their chests and stomachs that every curve of the breastbone and ribs stood out. When infants took on an unnatural, waxen appearance, their mothers saw it as an ominous sign—a death mask.

As difficult as these problems have been, successive El Niños now threaten to make things much worse. The 1998 El Niño drought was billed as the most severe of the twentieth century and affected some ten million people in the region. It reduced millions of acres to a dustbowl of bone-dry river beds and cracked mud. Bleached skulls of dead cattle dotted the landscape. Adding to the problem, high urban unemployment in cities like São Paulo blocked the possibility of migration for poor peasants. After suffering catastrophic crop losses, famished farmers stormed markets and warehouses. Desperate, they even attacked trucks carrying food aid.[1]

When the mayor of the small town of Ouriciri refused to attend a meeting to discuss the distribution and control of food assistance, saying he "was busy traveling," incensed activists passed a resolution demanding that food resources be channeled through them.[2] Taking matters into their own hands, people then raided a local market and made off with all of the produce. After the raid, the authorities arrested members of the Movimento Sem Terra (MST), the Landless Rural Workers' Movement.

Founded by displaced farmers and Indians in the mid-1980s, the MST criticized Brazil's rampant inequality and highly concentrated system of land ownership and called on the government to expropriate land for the poor. By the 1990s, the group had already carried out hundreds of land takeovers and at 500,000 members had created the largest rural movement in South America. Despite being stridently socialist in its rhetoric, the MST also acquired a high degree of marketing savvy, which allowed it to turn the issue of land reform into an urgent national debate. When the drought hit in 1998, the MST accused the federal government of adopting a superficial drought relief program that lacked structural reforms such as technology and credits. What's more, the authorities had paid little heed to the dire warnings from satellite and agricultural experts. "The government knew the drought was coming," said the editor of one local newspaper. "It was no mystery. But the government was not prepared. They took no preventive actions."[3] In light of the botched response to the drought, the MST claimed that peasants had little recourse other than to loot government food warehouses.

After the raid on the local food market in Ouriciri, the police picked up a young twenty-seven-year-old MST militant and held her in jail. The woman had given up her position as a school teacher to work as a literacy instructor in squatter settlements around Ouriciri, and she and other local residents had grown weary of inattentive local officials. Once the young woman was arrested, her mother and sixteen-month-old daughter camped out at the local jail where she was being held. The two were joined by dozens of other MST members. Though the judge barred any visitors, the young militant managed to smuggle out a note. The raid was just and necessary, she wrote. "All we want is freedom from hunger and my freedom to teach the children and the adults in the simple class rooms and schools that we have built ourselves."[4]

As the MST has found out, achieving progressive environmental reform can be challenging. President Fernando Henrique Cardoso criticized MST takeovers in the wake of the drought. Like Peru's Alan García, Cardoso jettisoned his former ideals in the interest of furthering his political career. After assuming office in 1994 the former Marxist sociologist did a 180-degree turn, adopting "neoliberal" market-driven policies, including the privatization of state industry, which wound up costing peasants more than a million jobs.

An MST critic, Cardoso remarked that "promoting looting assaults the interests of the people. It has only one meaning: It is disorder, it is a mess, it is posturing for the media."[5] Needless to say, Cardoso failed to take a satisfactory stand on climate change and drought in the northeast. During a meeting with Bill Clinton in 1997 Cardoso said, "We should reduce the greenhouse gases, but in such a way as to ensure that we're not harming the interests or the development of any country—the United States, Brazil, or developing countries."[6]

Cardoso's inaction came as a disappointment to many in the northeast. There, a "drought industry" managed by political bosses and landowners had emerged that kept peasants dependent on emergency food and resources in order to control their votes. Critics charged that by the time of the 1998 drought nothing had changed: During the relief effort the government provided inadequate and often bug-infested food aid to the people, and demanded local residents' support and votes in return. Cardoso was intimately tied up with the local bosses in the area, including those blamed for preserving the semi-feudal social and agricultural conditions that had resulted in drought and famine. In the northeast, an authoritarian system reigned supreme in which the landowners controlled access to water, agricultural technology, and other resources. Known as the "colonels," these oligarchs commanded local political machines and virtual private armies. Because their power depended on a vulnerable and submissive populace, the colonels resisted efforts to modernize farming and reclaim the land.

Unfortunately, the future is becoming increasingly bleak for MST activists and others who live in areas threatened by climate change and food scarcity. Scientists warn that global warming will accelerate desertification in the Brazilian northeast, spurring more poverty and emigration.[7] Currently, farmers in that region will go to any lengths to get potable water, even hitching up an oxcart and traveling to wells miles away from their homes. The trip can take hours, and at the end of the day the water can be reddish brown, muddy, and polluted. In really dry years, when there is no rain for months, wells and streams run dry, and crops refuse to grow. To the west in the Amazon agriculture will also become more fraught—warmer temperatures and changes in precipitation will affect plantation forestry and farming, particularly subsistence farming. Rainfall reductions in critical dry months will lead to more water evaporation from plant leaves. Pest infestation will increase, leading to reduced agricultural yields.

Officials are worried that Brazil might have to rethink its agricultural geography. The agricultural south could soon experience extreme weather conditions, including intense Indian summers and strong rains. Climate change could derail Brazil's plans to become an agricultural powerhouse as coffee, soy, corn, grain, and other top crops decline. Scientists say that soy could be the hardest hit, with land suitable for the crop's cultivation declining 20 percent by 2020 and 40 percent by 2070 under even the most optimistic scenarios. Only sugar cane, which thrives in high temperatures, will benefit from the weather changes: In a few decades the crop could actually double in size.[8]

Cassava, another traditional Brazilian crop, is in particular danger from the effects of climate change. Also known as tapioca, cassava was discovered and domesticated by Amazonian Indians. When the Portuguese arrived in Brazil in the early 1500s, they observed that cassava was the main staple for indigenous peoples. It is surprising, to put it mildly, that the Indians discovered that these tubers were edible in the first place—uncooked cassava contains potentially toxic concentrations of cyanogenic glucosides, a plant compound that contains sugar and produces cyanide. To detoxify the tubers, the Indians have to peel and grate them. They then put the pulp into long, thin cylinders called *tipitis*, and hang a heavy weight at the bottom, compressing the pulp and expelling the poisonous juice. Once they accomplish that, the Indians heat the starch settling out from the extracted juice upon a flat surface. Heating the detoxified starch separates the juices and causes individual grains to pop open and group together in round granules called tapioca. The Indians also roast the cassava pulp to produce *farinha de mandioca* (or manioc meal), a coarse flour. Today, *farinha* is as important a dietary staple as it was in pre-Columbian times and can be found on almost every Brazilian table.

Brazil can ill afford any problems with its cassava harvest. In the northeast, small-scale farmers rely on cassava to survive. Though the root does not provide complete nutrition, it contains high amounts of carbohydrates and starch and is very cheap. Cassava is very versatile; *farinha*, for example, can be used as a substitute for wheat flour when baking breads. In addition, the root can be boiled, baked, or fried like French fries. Even the unpeeled roots can be grated, dried, and later used as animal feed. Indeed, "it would be difficult to design a more appropriate food crop for the tropics than the cassava root."[9]

Latin American agriculture is in for some turbulent times. In the next few years, agricultural yields will plummet and more land will go without rain. Temperate zones will almost disappear in some countries, including Argentina and Chile. In the Amazon, and other tropical and subtropical areas, the growing season will be much shorter.

New York City is a long way from South America, but one man there has a unique perspective on what it takes to survive in Latin America during times of food scarcity. Miguel Pinedo-Vasquez is a research scholar at Columbia University's Center for Environmental Research and Conservation. A native son of the Peruvian Amazon, Vasquez was born in a small village located about four days by boat from Iquitos, the largest city in the Peruvian rainforest. Though for years he's been a resident of the East Village he still feels very connected to his Amazonian heritage. As a boy, Vasquez remembers how his mother and grandmother would celebrate when the floods would come to his village. "It was a blessing in many ways when the river would rise because it was a way of controlling the pests in the fields," he says. Flooding would kill off mice, rats, ants, and other pests, but unfortunately, the water would also wipe out local crops. In the Amazon, Vasquez explains, families traditionally had a lot of children to help out with the daily chores. He himself had seventeen brothers and sisters, of which five died. As a child, the boy had no other choice but to help out with the farming of bananas, cassava, corn, and beans. He also assisted his family with daily hunting and fishing.[10]

Since Vasquez left the Amazon a number of different animals have come under threat. One key food source for local residents is a fish called *arahuana*, which is traditionally hunted in late October or early November. But now that the biological cycle has been thrown off the season has started earlier.[11] Local residents have also had reason to be concerned about a local freshwater turtle, the *taricaya*, whose meat and eggs were a staple of the local diet. In years past the turtle was exported and sold in the United States for less than two dollars. Large-scale shipments were outlawed in 1975, however, when the *taricaya* was declared endangered. Peruvian authorities have tried to restore the animal population by constructing artificial beaches along riverbanks within the Pacaya Samiria Reserve. Scientists maintain a crucial egg incubation program and turtle young are later set free along the beaches. Despite such valiant efforts, climate change is endangering the *taricaya*'s recovery: Experts say that traditionally the turtle would lay its eggs in June or July but recently the season has moved up to August and September.[12]

Another way in which climate change is negatively affecting the turtle population of South America is by creating an unnatural shortage of males. What do increasing temperatures have to do with a dearth of males? In turtles, the determination of sex is very different than in mammals. The gender of a human child is determined by which chromosome, X or Y, it receives from its father, but most turtle species don't have these sex chromosomes. Instead, the sex of the offspring depends on egg temperature in the beginning of incubation. It's believed that cooler incubation produces more males and warmer incubation gives rise to more females. In Brazil's Mamirauá Reserve, the rising temperatures are now in the upper tier of incubation temperatures, and as a result fewer males are being hatched. A turtle population with a skewed sex ratio isn't good for the future because it means that a smaller group of males will wind up fathering the offspring. As a result, there will be less genetic diversity and less resistance to disease and climate. Unfortunately, it doesn't take much to skew sex ratios in some turtle populations: An increase of a mere few degrees during nesting season, if it persists over a generation or two, could throw off the gender balance enough to jeopardize a species.

In the Amazon residents face the decline of another traditional food source: tropical fruits. On my way to Iquitos recently I spotted a thick book in the Lima airport full of colorful photos of Amazonian fruits I had never set my eyes upon. A couple of days later in Iquitos, intent upon buying some of the fruits that I had seen in the photos, I made the rounds of some local stalls at the chaotic Belén market, where I picked up some cashew fruit as well as two less familiar fruits called *camu camu* and *aguaje,* also known as the moriche palm.

After returning to my hostel, the señora brightened when she saw my bag of *aguaje* and started to wax eloquent about her childhood in Iquitos. When she was a girl, she told me, she always used to eat aguaje from a tree located near her family's home. Heading to the kitchen, she soaked my aguaje in a pot of water. It wasn't until the next day that the fruit was soft enough to eat. The aguaje was difficult to peel and had a rather sour and buttery taste, but texture-wise, it reminded me of a baked potato. The señora suggested I sprinkle some sugar on the fruit, which improved matters somewhat.

Though little-known in the United States, aguaje is not only a highly nutritious fruit but also the most ecologically, economically, and socially important palm tree in Amazonia. Peru has five million total hectares of aguaje forest, one million of which is located near Iquitos in a nature reserve called Pacaya Samiria. The aguaje tree produces hundreds of oval-shaped fruit, which are covered in red wine-colored scales. Standing more than one hundred feet tall, the aguaje tree is an interesting sight: Its roots reach down into the waters of flooded swamps called *aguajales.* Two centuries ago, South American explorer Alexander von Humboldt (see Chapter Two) called the aguaje "the tree of life."

In Iquitos alone people eat about twenty tons of aguaje per day. Though aguaje is not processed on an industrial scale, the *aguajeras* (female fruit vendors) form an important part of the local economy and it's estimated that some 5,000 families derive economic benefit from the fruit.[13] No other Amazon fruit is sold in so many myriad ways: ripe, green, pulped into a local drink, and even made into popsicles, jelly, and yogurt. In downtown Iquitos, aguaje ice cream is a local favorite. The fruit's pulp is highly nutritious—its vitamin A content is five times higher than that of carrot and spinach. This high concentration of vitamin A makes aguaje an unparalleled dietary source for children and pregnant women because the fruit helps form and maintain healthy teeth, soft tissues and bones, mucous membranes, and skin. This invaluable fruit, however, has fallen under threat. Oil has polluted the Pacaya Samiria Reserve and killed off aguaje trees. Perhaps even more ominously, the aguaje ripening season has reportedly been delayed as a result of ongoing climate change.[14] Aguaje can become scarce, too, when the region gets hit by drought and the river dries up, making transportation difficult.

In addition to aguaje, the *camu camu* and *humarí* harvest has been delayed more than normal.[15] A small tree that thrives in wetlands, camu camu is particularly abundant in the Peruvian Amazon. The fruit is very high in vitamin C and is popular in drinks, popsicles, and candy. Both camu camu and aguaje serve as an important food source for local wildlife. Humarí is a purple-black, plum-shaped fruit with thick orange flesh. It has an odd, nutty flavor and is often eaten together with ripe bananas or even used as a substitute for butter at the kitchen table. But humarí production within the nearby Tamshiyacu Nature Reserve has fallen precipitously. "It was said that flooding and drought occurred in cycles," Juan Berchota of the Iquitos-based Institute for the Common Good remarks. "After the 1990s, this became more irregular. Farmers have a set calendar—they plant at such and such a time. But now there are no set seasons."[16] Concerned about falling yields, scientists at the Peruvian Institute for Amazon Research are now monitoring the situation.[17]

The problems that beset farmers in the Amazon are part of a larger dilemma facing people within the tropics worldwide. A recent article in the *Times* of London spelled out how India's mangoes are becoming scarcer and less sweet as a result of climate change interfering with harvests. In 2008 alone, three million tons of mangoes were wiped out by a harsh winter. Unseasonable deluges, meanwhile, swept key growing regions. Farmers are now lobbying the government to provide insurance to guard against the effects of unpredictable weather on their mango crops. Producers report that mangoes are losing their sweetness because hot, dry summer winds sweeping across northern and western India, which help to ripen the fruits, now fail to blow.

As global warming increasingly affects weather changes, the mango crop has been hit hard. In Uttar Pradesh, the second-largest mango-producing state, farmers reported that half the fruit harvest had been wiped out by storms within a two-month period. To add to the difficulties, unseasonable rain in the west encouraged pests, thus lowering output. At the Mango Mela, an agricultural fair held in the city of Bangalore, farmers featured only twenty varieties of mango in 2008, compared with more than one hundred in 2007. One farmer noted sadly that 75 percent of his crop had been wiped out by rain.[18]

So much for traditional food sources, but what about tropical cash crops? In my native Brooklyn I frequent a local café called the Cocoa Bar that specializes in coffee and chocolate. The place is quite popular and full of patrons on most any day of the week. Indeed, judging from the proliferation of espresso and chocolate bars throughout the country, I am not the only eager consumer of these two delicious tropical products. But what would Americans do if their precious caffeine and chocolate fix was cut off? Climate change could adversely affect production, resulting in the devastation of the plants that produce these substances and, subsequently, real economic problems for farmers in the South.

Consumers may not be aware that chocolate's long history in Latin America goes back thousands of years. Chocolate, which refers to any manufactured product made from the tropical cacao tree (*Theobroma cacao*), is formed by homogenizing cocoa powder with sugar, cocoa butter, and sometimes milk. Cocoa powder is produced by fermenting seeds from the cacao tree's pods, which are dried, roasted, and crushed. It is thought that the tree was domesticated by ancient civilizations, and cultivated by the Mayans, Aztecs, and Incas. For indigenous peoples, the tree's harvest was regarded as "a gift of the gods." Later, cacao beans were used to make the first stimulating drinks (meaning beverages containing caffeine and theobromine, alkaloids that affect the human body) encountered by Europeans.

Peruvians are proud of their country's long chocolate tradition and have played a key role in the development of the industry. In 1837, a young Italian candymaker set sail for South America. His name, Domenico Ghirardelli. After opening a confectionery store in Lima, Ghirardelli met an American named James Lick, who owned a cabinet shop right next to Ghirradelli's store. As it turned out, this was one of the happiest coincidences in chocolate's history. Lick, a businessman at heart, later returned to San Francisco, taking 600 pounds of Ghirardelli chocolate with him on his voyage. Once he landed in California, Lick wrote back to Ghirardelli in Lima, remarking, "This is a place of opportunity, and I would suggest that you bring yourself and some of your chocolates up here. I have sold the 600 pounds that I have brought and I feel there will be a great demand for it."[19] Encouraged by both the note and news of the California gold rush, the intrepid Ghirardelli set sail unaccompanied for the United States in 1849. Later, Ghirardelli opened the first candy store in San Francisco.

Today, cacao plays a key role in the fight against extreme poverty by creating employment and enabling small producers to enter into the world's prized consumption markets. However, cacao is now facing some environmental challenges. Peru in particular stands to be heavily impacted as the country sells tens of millions of dollars worth of the crop abroad. Unfortunately, cacao is particularly susceptible to disease because it is grown as a monoculture, the agricultural practice of growing a single crop over a wide area. A nasty fungus, frosty pod, has destroyed plantations in Colombia and Costa Rica.[20] Cacao's shallow roots have also responded poorly to droughts in recent years. Moreover, observers say that climate change is affecting cacao and coffee production, a significant economic blow to Peru. Normally, rain falls over the course of six months, but now that period has shortened to three or four months. Winter cold snaps have also been hitting the Peruvian countryside unseasonably, which causes the cacao crop to wither. [21]

As if concerns over cacao were not serious enough Peru must also confront problems with its coffee crop. Currently the Andean nation is the sixth-largest coffee producer in the world. Peru exports 95 percent of its coffee; 2007 sales amounted to $415 million. Most coffee growers are Indians who speak Spanish as a second language and live on small, remote farms without modern amenities. To harvest their crop, farmers move their hands methodically down the coffee branch, raking the red coffee cherries into the basket around their necks. The farmers then carry the ripe cherries to hand pulpers, who in turn process the coffee in wooden fermentation tanks. Finally, the farmers must haul their beans by foot or mule into the nearest town—a trip that can last anywhere from thirty minutes to eight hours.

Growing coffee is physically back-breaking work and is not very economically remunerative. The most a coffee-farming family can hope to earn is a paltry $1,200 a year, and in poorer areas some earn as little as $400. Peru's high jungle, the *selva alta*, which extends as high as 5,000 feet, is the center of the country's coffee production. Coffee farming there is a source of employment for a full 160,000 families, with another two million people along the production chain. In the coffee zone, social conditions are bleak: Transportation and roads are inadequate, people do not eat a balanced diet, and there are not enough schools.

Given their difficult economic position, coffee farmers can ill afford other problems. And yet, these migrant workers are becoming all too familiar with the effects of climate change. Coffee pickers, who move with the seasons in search of ripe coffee plants, must now begin the coffee season one month earlier as a result of the warmer temperatures. The coffee growers say seasons are changing so much that there is no longer any certainty that winter will fall in November, December, or March. This has resulted in a headache for coffee growers and caused total productive disorder. In addition, farmers report that high-altitude plants are maturing at times that are more typical of their lowland counterparts. Climatic unpredictability is eroding Peru's privileged position in the global coffee market. Traditionally coffee growers start to pick their crop in April, six months before the rest of the world. This gives Peru a unique competitive advantage. But if the seasons continue to move earlier, the country could lose this advantage. The long-term effects of climate change could also pose other unusual conditions for the Andean nation, potentially making new land available to farm coffee while at the same time exposing the crop to unusual precipitation and atypical humidity levels.

The problems confronted by Peruvian farmers are part of a larger, world-wide trend. The coffee crop has come under threat elsewhere in Latin America, and in some African countries, including Uganda, Kenya, and Tanzania, climate change could leave the land unsuitable for growing coffee entirely. In Costa Rica rising temperatures are pushing coffee growers to higher and higher elevations while, bizarrely, other areas in the country that were previously thought to be too chilly are now being converted into prime agricultural real estate. Currently, climate change is being blamed for a 2007 drought in Brazil and winds in Guatemala that slashed coffee harvests.[22] In Colombia, high in the Sierra Nevada Mountains, Arhuaco Indians know that something is amiss. Not only are their coffee plants flowering prematurely, but the coffee berries that dot the stems of their plants are smaller and weaker.[23]

When a full-blown food crisis hit the world in 2007, many in Latin America shook their heads in disbelief. Though the region had a rich agricultural tradition and several countries were among the world's largest food exporters, increased prices hit the poor particularly hard. The prospect of falling back into poverty was a particularly difficult pill to swallow given recent economic improvements. Poverty rates had been slashed in countries across Latin America and a new middle class had emerged in Brazil and Mexico, all thanks to consistent economic growth, low inflation, and government social spending. Latin Americans began to enjoy expanded access to consumer goods, home ownership and credit, and stable employment. All of these gains, however, stood to be erased when the economic slowdown hit the United States. Experts blamed the ensuing global crisis on a rise in oil and energy prices, which strained every phase of food production from fertilizer to tractors to transport. As the international price of such staples as maize, rice, vegetable oil, dairy products, soybeans, and wheat rose astronomically, the poor in developing countries around the world, including Latin America, were heavily impacted, as was the supply of ethanol and biodiesel for cars and trucks. In Haiti, starvation led to riots that toppled the government, and food shortages deepened the agony for cyclone survivors in Myanmar. In the end, the food crisis plunged 100 million people deeper into poverty and misery.

Yet other environmental factors, too, played a role in the crisis. According to experts, climate change-induced extreme weather including flood and drought contributed to an increase in food prices. From Africa to Australia to China, drought devastated local harvests. China was particularly hard hit with flooding, landslides, and mudslides killing 700 people across twenty-four provinces. In all, eighty-two million people were affected by the freakish weather that experts blamed on global warming. Hubei Province was battered by flooding from the Yangtze River while hailstorms and rain injured scores. In the mountainous Xinjiang region thousands were stranded as a result of landslides. In the south and east of the country, meanwhile, millions faced drought and acute drinking water shortages. Farmers, meanwhile, were left without irrigation water for their crops.

In Latin America, people wonder how climate change will affect the food supply and how they will cope. For centuries in Argentina some have believed that a local drink called yerba maté can help to reduce hunger and today the beverage is popular amongst the Argentine poor.[24] A drink similar to tea, maté is brewed from the dried leaves of a member of the holly family. Known as a stimulant, maté contains caffeine, vital vitamins and minerals, as well as other chemicals such as theobromine. A Danish study concluded that a compound containing maté, guaraná, and damiana slows the release of the contents of the stomach, resulting in an earlier sense of fullness when eating.

Hopefully, Argentines won't have to resort to drinking maté for the sake of reducing hunger, though recent developments do not bode well. In 2008, Argentines were hit by the food crisis as prices went through the roof. In short order, the poor lined up at church soup kitchens to receive free breakfasts. In Argentina, where officials had touted economic growth and the country's status as a top exporter of grain, vegetable oils, and beef, it felt as if the impossible had happened. Like other regimes around the world, the Argentine government slapped taxes on agricultural exports in the hope of lowering food prices at home. With world prices for wheat and soya beans at record levels, President Cristina Fernández de Kirchner argued that farmers ought to share their windfall profits with the rest of the country. Argentinean farmers blocked roads in protest.

As the food crisis worsened, leaders from the European Union and Latin America met in Lima. Participants stressed that they were deeply concerned by the impact of increased food prices. Environmental degradation and climate change, they agreed, seriously affected economic growth and struck the poor hardest. At the close of the conference the European Union agreed to a package of $183 million toward food aid world-wide. Nevertheless, outside the conference resentment seethed in Lima's poorest neighborhoods. There, residents felt that despite the EU–Latin American summit's claims, little was being done to address the day-to-day realities of poverty. South of the capitol in Villa Salvador, the 500,000 people living in rundown housing on shifting sand dunes grew especially indignant. There, taxi motorbikes sputter up and down unstable hills covered in garbage piles, while packs of dogs search through them for scraps. Residents survive by working small jobs that pay about ten dollars a day if they are lucky. It's a precarious existence, and skyrocketing prices for such staples as cooking oil and rice only made things worse. At the height of the crisis, some people at Villa Salvador reported that they couldn't even afford to buy vegetables.[25]

As a foreign visitor to Lima, it's easy to forget about places like Villa Salvador. Walking around the posh area of Miraflores, what is striking is the construction taking place on nearly every street. New hotels and restaurants have been sprouting up all over Lima and shopping centers proliferate. The city of eight million people is fast turning into a bustling metropolis, and has become the visible face of the boom that has made Peru South America's fastest growing economy. Amid all the plenty, however, poverty persists and food inflation has made things worse. Despite a six-year economic boom, benefits had barely trickled down to the poor, and many rarely had enough to eat. Knowing the situation could become politically dangerous, President García deployed the army to distribute direct aid to people living in wooden shacks on the outskirts of the city.

Peru was not alone in its troubles. In Brazil, the food crisis inflated the price of such staples as *fiejãozinho,* or beans and rice. Brazilian leaders, believing the crisis was limited to only a few items, blamed the shortages on rich countries' agricultural subsidies and support for corn-based biofuels. Despite such pronouncements, Brazil could do little in the short term to control commodity markets, which were driven by speculation on a number of factors, including skyrocketing demand in India and China.

In Mexico, the price of staples, including eggs, milk, and corn, shot up. Frightened for the future of their livelihoods, an alliance of farm groups sent a strongly worded message to President Felipe Calderón complaining of the high price of tortillas. When tens of thousands of people staged demonstrations to protest a 60 percent increase in the price of tortillas, an essential staple for the entire country, observers dubbed the conflict "the tortilla war."

Reeling from the food crisis, a number of Latin American and Caribbean countries sent representatives to an emergency conference in Nicaragua in 2008. A number of the region's left-leaning leaders were in attendance, including Evo Morales of Bolivia and Rafael Correa of Ecuador. Together, they criticized the world's developed countries for their role in creating the global food crisis. Cuban diplomat Laso Hernández blamed the catastrophe on climate change, high oil prices, "and the impact upon them of the military adventure in Iraq," as well as other factors.[26]

One of the more poignant moments of the conference came when Ralph Gonsalves, the prime minister of the Caribbean island of St. Vincent and the Grenadines, addressed the crowd. Unfortunately, he declared, U.S.-spurred global warming had increased ocean temperatures to such an extent that fish had been driven to ever deeper waters where they were more difficult to catch. For the tiny island nation climate change is hardly some kind of esoteric, theoretical concept. Indeed, scientists say that climate change is responsible for dramatic fish kills as higher-than-normal rainfall alters the salinity of the Caribbean Sea.[27] During the conference, Gonsalves remarked that hurricanes and tropical storms had wreaked havoc on lobster and shrimp populations inhabiting the sea beds, a phenomenon that the Caribbean statesman linked to global warming. All this had resulted in a fisheries-based food crisis for his compatriots.

Perhaps the greatest surprise came when Costa Rican president Óscar Arias, hardly a political radical, got his turn at the microphone. Arias said he couldn't understand how the United States would only offer $1 billion in food aid to the world even as it spent the same amount in half a week on the war in Iraq. He added that the Kyoto Protocol on climate change was a "great monument to hypocrisy" since the developed countries, "having polluted the planet in order to enrich themselves, are now asking us not to do so."[28]

CONCLUSION

I f we believe the experts, we don't have much time to spare: The longer we allow carbon emissions to increase the more challenging it will be to reduce them to within acceptable limits. In order to halt the planet's temperature from rising more than 2 degrees Celsius beyond pre-industrial levels, we must limit carbon emissions to no more than one trillion metric tons. Unfortunately, since the Industrial Revolution began, the world has exhausted more than half that amount. If we don't act fast we are likely to emit the rest of our precious budget within just a few decades, even if emissions are maintained at the current rate. We can ill afford to continue down this path—the result could be a cycle of ever-increasing and intense El Niños in the Amazon. It's a scenario to be avoided at all costs, since drought in the tropics will lead to forest fires and add to already worsening global warming. It would be great if there was a silver bullet that could save the Amazon, yet solving the tropical rainforest dilemma is an enormously complex social and political problem.

One thing is for sure, though—unless the Global North leads the way, poor countries are unlikely to take drastic steps to reduce their own emissions. Currently Brazil, China, and India are chief greenhouse gas emitters. In international negotiations, Brazil has refused to accept emission targets before the middle of the twenty-first century. The South American nation justifies this retrograde position by arguing that carbon dioxide, the most important greenhouse gas, stays in the atmosphere for more than a hundred years on average. In a sly move, Brazilian officials have argued that climate change is most crucially related to overall global temperature increases over time. Therefore, they conclude, earlier emissions should be weighed more heavily than yearly emissions in the present. At the current rate, Brazil could only be expected to meet emission targets by 2050, by which time the burden of responsibility for total emissions present in the atmosphere would be equivalent for the nations of the Global South and Global North.

Unfortunately, the Global North has yet to show the way out of our climate morass. Recent climate talks held in the German city of Bonn, which were designed to build momentum for the upcoming United Nations climate summit in Copenhagen, got bogged down with affluent nations resisting deep cuts in carbon emissions. While a deal might still be forged at Copenhagen, this will largely depend on what the United States comes forward with. So far Washington has sent out positive political signals that it wants to engage but hasn't been very specific about what it wants to put on the table. And here's the critical point: Even if negotiators cut a good deal in Copenhagen there's no guarantee that U.S. president Barack Obama can move intractable conservatives in Congress to approve an international treaty. In light of the political obstacles some have suggested that Obama should deal outside of the United Nations framework and concentrate instead on bilateral and multilateral negotiations with the world's major emitters, the most important of which is China.

On the other hand, even though the pace of international negotiations has been glacial and inadequate, it's imperative to hammer out a positive Reducing Emissions from Deforestation and Forest Degradation (REDD) plan at upcoming United Nations negotiations. Perhaps, REDD could include funds for reforestation so that we could make tropical forests part of the solution to global warming as opposed to an aggravating factor. In recent years scientists have tried to think of ways that reforestation can be a remedy for our climate ills. It turns out that when you plant trees in snowy areas like Canada and Siberia, the dark forest canopy absorbs sunlight that would have otherwise been reflected by the snow; the absorption creates more warming. Tropical forests, however, cool the planet more by absorbing water from the ground and helping to form thick cloud cover that reflects sunlight back into space. They also cool the earth through uptake of carbon. It's a win-win scenario: The more tropical trees we reforest the better off we'll be.

There's a catch though: Replanting tropical forests is prohibitively expensive, and in any case knowledge about how best to restore indigenous vegetation is lacking. Not surprisingly, seed dispersers such as birds and small mammals keep away from deforested land as these areas rarely offer much food or protection. Ingeniously, however, scientists have come up with a simple and cheap solution: They are now using bats to ramp up natural reforestation in the tropics. By installing man-made bat roosts in deforested areas, scientists hope to significantly increase seed dispersal of a wide variety of tropical plants. Bats could be the key to reforestation: They cover large distances during nightly foraging flights and, unlike other animals, are more willing to enter deforested areas. Because they eat fruit and nectar, the bats act as a crucial species for seed dispersal and flower pollination. The bats can use the roosts, which are simply built boxes, for many years. "We hope that this cheap and easy to use method will be applied in many parts of the tropics in the near future, and that bats will be 'employed' as efficient agents of reforestation," remarks one researcher. Shunned for millennia, these nocturnal creatures may hold the key toward ameliorating deforestation and battling climate change.

One of the thorniest and most complicated issues that the world will have to figure out is land use in the tropical rainforest. At its root, deforestation is linked to deep-seated social inequities and political injustices that have been inherent in Brazilian society. Veterans of the Landless Rural Workers' Movement (MST) who have been carrying out land occupations for years say land concentration and climate change are inextricably linked. In the first place, corporate conversion of forest for monoculture results in carbon emissions. Further, as agribusiness transports its goods all over the world, it increases emissions even more through use of fossil fuels. To top it off, industrial agriculture relies on fertilizers and equipment, such as tractors, that contribute to climate change. Agribusiness is environmental suicide, the MST argues, because such an approach effectively shunts the rural poor aside in the name of promoting so-called development. MST activists argue that we should encourage small-scale organic farming, local food consumption, and the eradication of nitrogen fertilizers. Brazil must completely dismantle agribusiness and make greater strides toward meaningful land reform, they argue. Reluctant to alter Brazil's skewed landholding patterns, the authorities have resettled many farmers on state-owned plots within the Amazon. That's a recipe for disaster, says the MST—the first priority should be agrarian reform carried out on existing agricultural land, not an expansion of the lawless agricultural frontier into environmentally sensitive areas.

Instead of funding agribusiness, the MST advocates small-scale farming and "agro-ecology" designed to protect and improve productive quality of the land. Landless peasants also push for agro-forestry, which combines agriculture and forestry in an effort to encourage sustainable land use and eliminate indiscriminate clearing of rainforest. By incorporating age-old agricultural tools of indigenous peoples—such as *terra preta* soil, which absorbs carbon dioxide—we can improve soil quality, support agriculture on deforested areas, and even fight global warming in the process.

Could tropical products such as chocolate be the silver bullet for our climate woes? While many would argue that chocolate is good for the soul, some think that it might also help the environment. In Eastern Brazil, farmers are doing their utmost to revive cacao production. While it used to be a huge industry in the region, over the past twenty years cacao has fallen on hard times as a result of plant disease and low prices on the world market. Squeezed by environmental and economic woes, farmers burnt the forest for farmland or pasture. Now, however, they are incorporating cacao trees into the rainforest. The farmers save the forest by promoting a method called *cabruca*—they cut down a few tall rainforest trees and plant medium-sized cacao trees underneath. Everywhere you look something is growing in the rainforest and the place is loud with the calls of tropical birds.

Provided that climate change and drought do not undermine *cabruca* farms, cacao could be a good option for rural inhabitants of the rainforest. In the *cabruca* forest it's moist, shady, and cool, and the ground is covered by a thick layer of composting leaves. In recent years there's been growing demand for environmentally friendly chocolate, and some cacao farmers have gotten a premium for their crop. All well enough, but some researchers have even more ambitious plans: Agronomists working with chocolate manufacturer Mars and Brazil's national chocolate research institute have been conducting research on land cleared of its forest cover. The soil initially was hard and compact and only weeds thrived there. Farmers arrived and planted some rubber trees as well as corn, beans, melons, squashes, and bananas. The soil recovered and became moist and crumbly. Eventually, scientists hope the soil will be rich enough to support profitable and environmentally sustainable cacao trees.

Although preserving or replanting forest will not fix global warming, it's a step in the right direction. To be sure, the international community can't dictate the terms of Brazilian land use, but it can opt to fund certain types of economic initiatives over others. For far too long the World Bank and the Inter-American Development Bank (IADB) have been supporting climate-unfriendly industries in the Amazon. For that reason, REDD should not make use of these large financial institutions when allocating vitally important adaptation funds. Fundamentally, we need a sea shift from agribusiness to sustainable forestry and the like, and international capital is going to have to radically rethink its entire *modus operandi*. Instead of funding Blairo Maggi and destructive cattle farming, the international community could support local cacao farmers, for example. Cacao farmers don't need any encouragement: They're already hoping to cash in on carbon credits for cultivating *cabruca* chocolate. Indeed, Mars, Inc. and the World Agro-forestry Center, based in Nairobi, Kenya, are studying how carbon storage can be measured on *cabruca*-type farms.[1]

There's no shortage of other diverse tropical products that could be exploited for commercial use, and donor countries in the Global North should be talking about how to fund them through REDD. Take, for example, Brazil nut and rubber trees, not to mention a range of fruits including *aguaje* or *camu camu*. Fruit in particular has great unrealized potential: Of the 3,000 rainforest fruits, only 200 are commonly consumed. By cultivating more fruit we could provide meaningful employment for local people as well as preserve the rainforest as an important carbon sink. Indeed, the swamps where the *aguaje* tree grows act as great carbon storehouses: It's been found that they store more than 600 tons of carbon dioxide per hectare. That's three to five times more than any other tropical ecosystem.[2] Provided that climate change does not throw off the fruiting season, *aguaje* could provide a highly nutritious staple for local inhabitants while simultaneously providing important environmental benefits for the world.

If we wish to keep our forests intact and maintain our climate equilibrium then we must consider other viable economic alternatives for indigenous people. Currently just 10 percent of natural food colorants are derived from rainforest products, yet production of rainforest colorings could be vastly expanded. Brazil could also promote an additional underutilized activity: medicinal drugs produced from rainforest plants. The drugs could be derived from bark, leaves, roots, and other plant parts. There are yet other economic options on the table. Through carefully planned eco-tourism, forest dwellers can gain employment as wildlife guides and park rangers. In the absence of effective environmental oversight and an uneven performance at best by agencies such as Ibama, indigenous guides are ideally situated to report environmental abuse within parks or Indian reserves.

In other parts of the Amazon the international community could play a vital role in arresting climate change and promoting alternative development if it chose to do so. Like Peru, Ecuador has developed its own share of Amazon rainforest for oil development. When environmentalists and Indians grew concerned that major oil fields might be developed in a lush area of rainforest called Yasuní National Park, which was declared a UNESCO biosphere reserve in 1989, Ecuadoran president Rafael Correa asked the international community to compensate Ecuador for half of the forecasted lost oil revenue in exchange for not carrying out petroleum development in the area.

"Ecuador doesn't ask for charity," remarked President Correa, "but does ask that the international community share in the sacrifice and compensates us with at least half of what our country would receive, in recognition of the environmental benefits that would be generated by keeping this oil underground." Environmentalists praised Correa for his "debt-for-carbon swap," which they said would result in the sequestration of up to half a billion tons of carbon dioxide. Two years after making his offer, however, Correa grew incensed at the lack of international response. The Global North, Correa declared, always talked about the need to preserve the Amazon but when push came to shove refused to come up with any funding. "Up to now there hasn't been two cents" from the international community, Correa lamented.[3]

For affluent nations, promoting smaller-scale development and avoiding large mega-projects in the Amazon would be a welcome step in the right direction vis-à-vis climate change. Rich countries, however, must also take decisive action on other fronts, such as moving beyond the fossil fuel paradigm and funding alternative energy initiatives, such as wind and solar, in the Global South. Yet, affluent nations should work toward even more radical technological breakthroughs in an effort to take some of the climate pressure off people living in the tropics. Observing the pitiful lack of international progress on climate change, some scientists now advocate more radical "geo-engineering" proposals to solve our global warming dilemma. Some of the ideas sound like something out of a Hollywood science fiction movie. Take, for example, the notion of mirrors that could reflect sunlight back into space. Mirrors, scientists say, could be put into space or reflecting film could be placed in the desert. Another option might be to construct floating white plastic islands in the ocean that could mimic the effect of reflective sea ice. One scientist from Columbia University has designed a "synthetic tree" that can collect carbon from the air. The device looks less like a tree and more like a small building, and can collect carbon about 1,000 times more rapidly than a normal tree. The synthetic tree captures carbon in a chamber and then compresses and stores it in liquid form for sequestration.

Nobel prize-winning Dutch scientist Paul Crutzen, who has done much to advance our understanding of nitrous oxide and its connection to climate change, proposes an even more far-out idea: launching balloons or rockets to disperse one million tons of the common pollutant sulfur into the upper atmosphere every year. The sulfur particles, Crutzen says, could help to reflect sunlight and heat back into space, thereby helping to cool the planet. Crutzen's plan is modeled in part upon a volcanic eruption that occurred on Mount Pinatubo, Philippines, in 1991. During and after the event, thousands of tons of sulfur were spewed into the atmosphere and global temperatures fell by 0.5 C on average in 1992. The cooling occurred, scientists say, because the sulfate particles acted as tiny mirrors that prevented some incoming light from reaching the surface of the Earth. Crutzen frankly admits that there could be unwanted side effects to his plan, including an increase in the destruction of the ozone layer and, bizarrely, a whitening of the sky. On the plus side, Crutzen adds, the particles would make sunsets and sunrises more spectacular. Some researchers in England comment that of all geo-engineering proposals on the table, Crutzen's plan has the greatest potential to cool the climate by 2050. On the other hand, they say, Crutzen's initiative carries risks and they warn that "geo-engineering alone cannot solve the climate problem."[4]

For us to avert a true environmental calamity the Global North must work collectively with the peoples of the south. There will not be any one approach on its own that will solve our problem; rather we must employ a variety of political, social, and technological strategies simultaneously. Above all we must approach climate change with much more of a sense of urgency than we have heretofore demonstrated. In the Global South they're much more aware of the need for drastic and prompt action. For centuries the spiritual elders, or *mamos,* of the Arhuaco tribe of Colombia have carried out monthly rituals in sacred sites throughout the mountains. They call these sites "the Heart of the World" and the rituals are designed to ensure that the Earth is maintained in geo-spiritual balance. Since the 1980s, however, the *mamos* have observed rapid changes in the Heart of the World, including lower moisture levels and changing patterns in bird and butterfly migration.

Alarmed, the elders have opted to share their observation with the rest of the Arhuaco tribe as well as outsiders, the so-called "Younger Brothers." These outsiders, the Arhuaco believe, have come to "the Heart of the World" and cut out the Mother's heart. The outsiders cut down trees that hold the earth in place, also destroying bird habitat. They pollute the water with chemicals from mining and make drugs from traditional plants, such as the sacred coca leaf. Even more seriously, say the *mamos,* the Younger Brothers have changed the whole earth and the

Mother is getting warmer. The rain comes later and falls harder, and the bees are disappearing, affecting the flowering of coffee and other plants. When asked how he knew there were fewer bees, a *mamo* remarked "I can hear them. Their sound has lessened. It is all happening very quickly. First you [the Younger Brothers] took our gold. Then you took our land. Now you are taking the water and the air itself. The Younger Brothers are waging a war on the earth and it must stop!"[5]

NOTES

INTRODUCTION

1. Since this book went to press, however, David Fogarty asserts in a *Scientific American* article "Forest Carbon Scheme Hopes for Green Light in Copenhagen," that deforestation results in a lower figure of 12% of greenhouse gas emissions.
2. Steve Connor, "Revenge of the rainforest," *Independent,* March 6, 2009, http://www.independent.co.uk/environment/climate-change/revenge-of-the-rainforest-1638524.html (accessed August 9, 2009).
3. "Rain forest threat isn't close to abating," *San Francisco Chronicle,* February 29, 2008.
4. Trish Anderton, "VOA News: Scientists fear Amazon may face early destruction," U.S. Fed News (a news service providing transcripts of official U.S. government hearings, speeches and press conferences and available through Factiva data base), December 12, 2007.

CHAPTER 1

1. "Central Andes Water: Future Shortages Feared as Glaciers Melt," Latin American Special Reports (part of LatinNews and the Intelligence Research Group, available on Factiva and on the web at http://www.latinnews.com/), October 25, 2005.
2. Interview with Marco Zapata, Huaraz, October 10, 2008.
3. Interview with Erick Meneses, Lima, October 7, 2008.
4. Leslie Josephs, "Calentamiento global derrite glaciares andinos de Perú," AP Spanish Worldstream, February 11, 2007.
5. "Peruvian Glacier May Vanish In 5 Years," Science Daily, February 18, 2007, http://www.sciencedaily.com/releases/2007/02/070215181454.htm (accessed July 13, 2009); "Ohio Professor Wins Science Award," Associated Press, May 31, 2007; J. Madeleine Nash, "Chronicling the ice: Long before global warming became a cause celebre, Lonnie Thompson was extracting climate secrets from ancient glaciers. He finds the problem is even more profound than you might have thought," *Smithsonian,* July 1, 2007; Doug Struck, "Global Warming Threatens Water Supplies—Some Parts of World Already Feel Stress as Droughts Persist," *The Washington Post,* August 21, 2007.
6. "Huascarán National Park," Unesco World Heritage Centre website, http://whc.unesco.org/en/list/333 (accessed December 3, 2008).

7. "Humachuco—Culture," Mountain Institute website, http://www.mountain.org/work/andes/tourism/humachuco03.cfm (accessed December 3, 2008).

8. "El Oso de Anteojos," Instituto Machu Picchu, http://www.imapi.org.pe/7boletin01.htm (accessed December 3, 2008).

9. Max Milligan, "Inca Spots," *The Guardian,* September 29, 2001, http://www.guardian.co.uk/travel/2001/sep/29/peru.wildlifeholidays.guardiansaturdaytravelsection (accessed June 27, 2009).

10. Interview with Manuel Glave, Lima, October 2, 2008.

11. "Machu Picchu Historic Sanctuary," World Monuments Fund (private organization working to preserve historic monuments), Web site, http://wmf.org/watch2008/watch.php?id=S407 (accessed July 13, 2009).

12. Interview with Cesar Moran Cahusac, executive director of Amazon Conservation Association, Washington, D.C., August 25, 2008; CONAM, *Manos a La Obra, El Cambio Climático en el Desarrollo Sostenible del Peru,* undated book, page 15; Abraham Lama, "Glacial Snow Disappearing in the Andes," Interpress Service, 1999, http://www.converge.org.nz/lac/articles/news990724e.htm (accessed December 3, 2008); interview with Manuel Glave, Lima, October 2, 2008.

13. Interview with Glave, Lima, October 2, 2008.

14. Interview with Zapata, Huaraz, October 10, 2008.

15. "Early Climate Change Victim: Andes Water," *Miami Herald,* November 23, 2007, posted on *USA Today* website, http://www.usatoday.com/weather/climate/2007-11-23-andes_N.htm (accessed December 3, 2008).

16. Interview with Zapata, Huaraz, October 10, 2008.

17. "Early climate change victim: Andes water," *Miami Herald,* November 23, 2007, posted at *USA Today* website, http://www.usatoday.com/weather/climate/2007-11-23-andes_N.htm (accessed December 3, 2008).

18. Interview with Zapata, Huaraz, October 10, 2008.

19. Interview with Laura Avellaneda, Lima, October 1, 2008.

20. "Perú busca defenderse de la amenaza del calentamiento global," Agence France Presse, May 31, 2007.

21. "García plantea desalinizar el mar como objetivo nacional de Perú," Reuters–Noticias Latinoamericanas, March 11, 2008.

22. Leslie Josephs, "Calentamiento global derrite glaciares andinos de Perú," AP Spanish Worldstream, February 11, 2007.

23. "García plantea desalinizar el mar," Reuters–Noticias Latinoamericanas, March 11, 2008.

24. *Amigos de la tierra internacional, Cambio climático,* Amsterdam, 2007, 24.

25. Joanne Silberner, "In Highland Peru, a Culture Confronts Blight," National Public Radio, *All Things Considered,* http://www.npr.org/templates/story/story.php?storyId=87811933 (accessed December 3, 2008).

26. Ibid.

27. Eliza Barclay, "Warming Andes Stymies Peruvian Potato Farmers," *San Francisco Chronicle,* October 5, 2008, http://www.sfgate.com/cgi-bin/article.cgi?f=/c/a/2008/10/04/MNTK133AIF.DTL (accessed December 3, 2008).

28. Ibid.

29. Milagros Salazar, "PERU: Preserving the Potato in Its Birthplace," Inter Press Service, http://ipsnews.net/news.asp?idnews=42161 (accessed July 13, 2009)

30. Silberner, "In Highland Peru, a Culture Confronts Blight," NPR *All Things Considered,* http://www.npr.org/templates/story/story.php?storyId=87811933 (accessed July 13, 2009).

31. Salazar, "PERU: Preserving the Potato in Its Birthplace," Inter Press Service, http://ipsnews.net/news.asp?idnews=42161 (accessed July 13, 2009).

32. Silberner, "In Highland Peru, a Culture Confronts Blight," NPR *All Things Considered,* http://www.npr.org/templates/story/story.php?storyId=87811933 (accessed July 13, 2009).

33. Salazar, "PERU: Preserving the Potato in Its Birthplace," Inter Press Service, http://ipsnews.net/news.asp?idnews=42161 (accessed July 13, 2009).

34. Silberner, "In Highland Peru, a Culture Confronts Blight," NPR *All Things Considered,* http://www.npr.org/templates/story/story.php?storyId=87811933 (accessed July 13, 2009)

35. Amigos de la tierra internacional, *Cambio climático,* 24.

36. Amigos de la tierra internacional, *Cambio climático,* 25.

37. Daniel Howden, "Christmas Appeal: Simple Measures that Help in Extreme Temperatures," *Independent,* December 17, 2005, http://www.independent.co.uk/news/world/politics/christmas-appeal-simple-measures-that-help-in-extreme-temperatures–519774.html (accessed December 3, 2008).

38. "Developing Countries Brace for Climate Change Impact, Peru: Retreating Glacier," World Bank Web site, http://web.worldbank.org/WBSITE/EXTERNAL/NEWS/0,contentMDK:21578916~pagePK:64257043~piPK:437376~theSitePK:4607,00.html (accessed December 3, 2008).

39. Amigos de la tierra internacional (Friends of the Earth, a network of environmentalists), *Cambio climatico* (paperback book), Amsterdam, 2007, page 26.

40. Howden, "Christmas Appeal," *Independent,* December 17, 2005, http://www.independent.co.uk/news/world/politics/christmas-appeal-simple-measures-that-help-in-extreme-temperatures–519774.html (accessed July 13, 2009).

41. "Surviving the Friaje, Freak freeze conditions in Peru," http://practicalaction.org/?id=climatechange_friaje (accessed July 13, 2009).

42. "Clarity Brought to Earth's Cloud Forests," United Nations Environment Program, February 9, 2004, http://new.unep.org/Documents.Multilingual/Default_pub.asp?DocumentID=384&ArticleID=4406&l=en (accessed July 13, 2009).

43. Gerard Wynn, "Climate Change Threatens Latam Water Supply—Wbank," Reuters News, July 20, 2007; "Clarity Brought to Earth's Cloud Forests," United Nations Environment Program, February 9, 2004, http://new.unep.org/Documents.Multilingual/Default_pub.asp?DocumentID=384&ArticleID=4406&l=en (accessed July 13, 2009).

44. "Clarity Brought to Earth's Cloud Forests," United Nations Environment Program, February 9, 2004, http://new.unep.org/Documents.Multilingual/Default_pub.asp?DocumentID=384&ArticleID=4406&l=en (accessed July 13, 2009).

45. John Roach, "Cloud Forests Fading in the Mist," National Geographic News, August 13, 2001, http://news.nationalgeographic.com/news/2001/08/0813_cloudforest.html (accessed July 13, 2009).

46. "Clarity Brought to Earth's Cloud Forests," United Nations Environment Program, February 9, 2004, http://new.unep.org/Documents.Multilingual/Default_pub.asp?DocumentID=384&ArticleID=4406&l=en (accessed July 13, 2009).

47. Stephen Leahy, "Climate Change: Will Forests Adapt to a Warmer World?" Inter Press Service, November 20, 2006, http://www.ipsnews.net/news.asp?idnews=35548 (accessed July 13, 2009).

48. "Scientists Discover New Species Of Distinctive Cloud-forest Rodent," *Science Daily,* January 25, 2007, http://www.sciencedaily.com/releases/2007/01/070124175503.htm (accessed July 13, 2009).

49. "Clarity Brought to Earth's Cloud Forests," United Nations Environment Program, February 9, 2004, http://new.unep.org/Documents.Multilingual/Default_pub.asp?DocumentID=384&ArticleID=4406&l=en (accessed July 13, 2009).

50. Jeremy Hance, "Smallest Andean Frog Discovered in Cloud Forests of Peru," mongabay.com, March 18, 2009, http://news.mongabay.com/2009/0317-hance_andeanfrog.html (accessed July 13, 2009).

51. Leahy, "Climate Change: Will Forests Adapt to a Warmer World?" Inter Press Service, November 20, http://www.ipsnews.net/news.asp?idnews=35548 (accessed July 13, 2009).

52. "UN Climate Fund to Help Poor Countries," *Sydney Morning Herald,* December 13, 2008, http://news.smh.com.au/world/un-climate-fund-to-help-poor-countries-20081213-6xtq.html (accessed July 26, 2009).

CHAPTER 2

1. John Vidal, "Climate Scientists Warn of Wild Weather in the Year Ahead as El Niño begins," *Guardian,* July 13, 2009, http://www.guardian.co.uk/environment/2009/jul/13/el-nino-climate-change (accessed July 20, 2009); David Fogarty, "Emerging El Niño Set to Drive up Carbon Emissions," Reuters, July 7, 2009, http://in.reuters.com/article/environmentNews/idINTRE56604320090707 (accessed July 20, 2009).

2. "New Type of El Niño Could Mean More Hurricanes Make Landfall," Physorg.com, July 2, 2009, http://www.physorg.com/news165763631.html (accessed July 20, 2009).

3. David Fogarty, "El Niño Study Challenges Global Warming Intensity Link," Reuters, March 24, 2009, http://in.reuters.com/article/worldNews/idINIndia-38669820090324?sp=true (accessed July 20, 2009); Marcela Valente, "Climate Change Fuels Spread of Dengue Fever," Inter Press Service, March 19, 2007, http://www.ipsnews.net/news.asp?idnews=36994 (accessed July 20, 2009); "Penguins in Trouble from Global Warming," *Economic Times,* July 1, 2008, http://economictimes.indiatimes.com/Earth/Global_Warming/Penguins_in_trouble_from_global_warming/articleshow/3183404.cms (accessed July 20, 2009).

4. Fogarty, "Emerging El Niño Set to Drive Up Carbon Emissions," Reuters, July 7, 2009, http://in.reuters.com/article/environmentNews/idINTRE56604320090707 (accessed July 20, 2009).

5. Ian Sample, "Global meltdown: scientists isolate areas most at risk of climate change," *The Guardian,* February 5, 2008, http://www.guardian.co.uk/environment/2008/feb/05/climatechange?gusrc=rss&feed=11 (accessed July 20, 2009); Vidal, "Climate scientists warn of wild weather in the year ahead as El Niño begins," *Guardian,* July 13, 2009, http://www.guardian.co.uk/environment/2009/jul/13/el-nino-climate-change (accessed July 20, 2009).

6. Interview with Philip Fearnside, Manaus, October 24, 2008.

7. Philip M. Fearnside, "The Fractured Landscape," *American Prospect,* August 13, 2007, http://www.prospect.org/cs/articles?article=the_fractured_landscape (accessed July 20, 2009); "Climate Change 'to Hit Pacific Islands,'" *The Age,* April 9, 2007, http://www.theage.com.au/news/World/Climate-change-to-hit-Pacific-islands/2007/04/09/1175971008082.html (accessed July 20, 2009).

8. Interview with Philip Fearnside, Manaus, October 24, 2008.

9. Curt Suplee, "El Niño/La Niña, Nature's Vicious Cycle," *National Geographic*, http://www.nationalgeographic.com/elnino/mainpage.html (accessed July 20, 2009).

10. "Fish Story Linked to Climate Cycle," NASA Earth Observatory website, June 18, 2004, http://earthobservatory.nasa.gov/Newsroom/view.php?id=24955 (accessed July 20, 2009).

11. Cynthia Rosenzweig and Daniel Hillel, *Climate Variability and the Global Harvest: Impacts of El Niño and Other Oscillations on Agro-Ecosystems* (New York: Oxford University Press, 2008), 94, Google Books, http://books.google.com/books?id= 8-UbmSxLWlUC&pg=PA94&lpg=PA94&dq=el+nino+1972+fish+peru&source =bl&ots=453ZQ80ARS&sig=6IL0zXpQamP_HpsfN-nYM1KNB4I&hl=en &ei=N4hfSoPQGY-OMYvYua4C&sa=X&oi=book_result&ct=result&res num=1 (accessed July 20, 2009).

12. University of Washington Department of Atmospheric Sciences, "Reports to the Nation On Our Changing Planet," http://www.atmos.washington.edu/gcg/ RTN/rtnt.html (accessed December 4, 2008).

13. Joanne Silberner, "Climate Connections," NPR *All Things Considered*, February 25, 2008, http://www.npr.org/templates/story/story.php?storyId=19344123 (accessed December 4, 2008).

14. "Global: Rich Must Pay Climate Change Health Costs," IRIN, April 7, 2008, http://www.alertnet.org/thenews/newsdesk/IRIN/efb10252fe24c9c96388602d4 3d3fca2.htm.

15. James Brooke, "Cholera Kills 1,100 in Peru," *New York Times*, April 19, 1991, http://query.nytimes.com/gst/fullpage.html?res=9D0CE7D9173EF93AA2 5757C0A967958260&sec=&spon=&pagewanted=1 (accessed December 13, 2008).

16. Ibid.

17. "Global: Rich Must Pay Climate Change Health Costs," IRIN, April 7, 2008, http://www.alertnet.org/thenews/newsdesk/IRIN/efb10252fe24c9c96388602d4 3d3fca2.htm (accessed July 27, 2009).

18. Suplee, "El Niño/La Niña, Nature's Vicious Cycle," *National Geographic*, http:// www.nationalgeographic.com/elnino/mainpage.html (accessed July 20, 2009).

19. Clifford Krauss, "Fujimori's Burden in Peru: The Magic's Missing," *New York Times*, January 14, 1999, http://query.nytimes.com/gst/fullpage.html?res=950CE ED71231F937A25752C0A96F958260 (accessed December 6, 2008).

20. "El Nino Battering S. America Economies—Peru Bears the Brunt of Chaotic Weather," *Seattle Times*, March 5, 1998, http://community.seattletimes.nwsource .com/archive/?date=19980305&slug=2737937 (accessed December 6, 2008).

21. Krauss, "Fujimori's Burden in Peru," *New York Times*, January 14, 1999, http:// query.nytimes.com/gst/fullpage.html?res=950CEED71231F937A25752C0A96 F958260 (accessed December 6, 2008).

22. Brooke, "Cholera Kills 1,100 in Peru," *New York Times*, April 19, 1991, http:// query.nytimes.com/gst/fullpage.html?res=9D0CE7D9173EF93AA25757C0A9 67958260&sec=&spon=&pagewanted=1 (accessed December 13, 2008).

23. "Global: Rich must pay climate change health costs," IRIN, April 7, 2008, http:// www.alertnet.org/thenews/newsdesk/IRIN/efb10252fe24c9c96388602d43d3fc a2.htm (accessed July 27, 2009).

24. Suplee, "El Niño/La Niña, Nature's Vicious Cycle," *National Geographic*, http:// www.nationalgeographic.com/elnino/mainpage.html (accessed July 20, 2009).

25. Sebastian Rotella, "Fujimori Takes on El Nino," March 2, 1998, *Los Angeles Times*, http://articles.latimes.com/1998/mar/02/news/mn-24629 (accessed December 6, 2008).

26. Richard Collings, "Despatches," BBC, February 2, 1998, http://news.bbc.co.uk/2/hi/despatches/52676.stm (accessed December 6, 2008).

27. "El Niño in Peru: A New Lake and a Wake of Destruction," CNN, http://www.cnn.com/SPECIALS/el.nino/peru/ (accessed December 6, 2008).

28. Krauss, "Fujimori's Burden in Peru: The Magic's Missing," *New York Times*, January 14, 1999, http://query.nytimes.com/gst/fullpage.html?res=950CEED712 31F937A25752C0A96F958260 (accessed December 6, 2008).

29. Tim Radford and Paul Brown, "This Is What We Know About Global Warming . . . so Why Haven't We Done Anything About It Yet?," *The Guardian*, April 29, 2004, http://www.guardian.co.uk/science/2004/apr/29/environment .climatechangeenvironment (accessed January 12, 2009).

30. "Changes in Climate, Changes in Lives, How Climate Change Already Affects Brazil," Greenpeace website, http://www.greenpeace.org.br/clima/pdf/catalogue _climate.pdf (accessed January 12, 2009).

31. "Storms and Hurricanes Return with More Force, GEO Year Book 2004/5," United Nations Environment Program, http://www.unep.org/geo/yearbook/yb2004/045.htm (accessed January 12, 2009).

32. "Changes in Climate," Greenpeace, http://www.greenpeace.org.br/clima/pdf/catalogue_climate.pdf (accessed January 12, 2009).

33. Adam Morton, "Winds of Change Blow Ill for South Atlantic Storms," *The Age*, September 26, 2005, http://www.theage.com.au/news/national/winds-of-change-blow-ill-for-south-atlantic-storms/2005/09/25/1127586746538.html (accessed January 12, 2009).

34. Mike Davis, "The Other Hurricane, Commentary: Has the Age of Chaos begun?" *Mother Jones*, October 7, 2005, http://www.motherjones.com/commentary/columns/2005/10/other_hurricane.html (accessed January 12, 2009).

35. Morton, "Winds of Change," *The Age*, September 26, 2005, http://www.theage .com.au/news/national/winds-of-change-blow-ill-for-south-atlantic-storms/2005/09/25/1127586746538.html (accessed January 12, 2009).

36. "Climate Change Summit Opens with Call for Global Solidarity," *Economic Times*, December 2, 2008, http://economictimes.indiatimes.com/ET_Cetera/Climate_change_meet_for_global_solidarity/articleshow/3782978.cms (accessed July 27, 2009).

CHAPTER 3

1. Indira A.R. Lakshmanan, "Amazon Burning," *Boston Globe*, November 27, 2006, http://www.boston.com/news/globe/health_science/articles/2006/11/27/amazon_burning/ (accessed June 28, 2009); "Environment: Climate Change Spurs Lightning Strikes and Death," Inter Press Service, April 4, 2001, http://www.highbeam.com/doc/1P1-43284163.html (accessed July 23, 2009); "Rain Forest Threat Isn't Close to Abating," *San Francisco Chronicle*, February 29, 2008.

2. Trish Anderton, "VOA News: Scientists Fear Amazon May Face Early Destruction," US Fed News, December 12, 2007.

3. John Dorfman, "The Amazon Trail," *Discover* magazine, May 1, 2002, http://discovermagazine.com/2002/may/featamazon (accessed February 19, 2009); Fred Pearce, "Virginity Lost," *Conservation* 8, no. 1 (January–March 2007), http://

www.conservationmagazine.org/articles/v8n1/virginity-lost/ (accessed February 19, 2009).

4. Rhett Butler, "Amazon Stonehenge," mongabay.com, May 14, 2006, http://news .mongabay.com/2006/0514-amazon.html (accessed January 17, 2009); N. P. S. Falcão, C. R. Clement, S.M. Tsai, and N. B. Comerford, "Pedology, Fertility, and Biology of Central Amazonian Dark Earths," in William Woods, *Amazonian Dark Earths* (New York: Springer, 2008), 214, Google books, http://books.google .com/books?id=46bEXHqB_vYC&pg=PA127&lpg=PA127&dq=marajo+terra+ preta&source=bl&ots=DViGPwhK4l&sig=NsHuF3J20DAmOcpjT_89PI wSIEc&hl=en&ei=eUieSY7nBN-BtwfRwMGODQ&sa=X&oi=book_ result&resnum=9&ct=result#v=onepage&q=marajo%20terra%20preta&f=false (accessed October 2, 2009).

5. Pearce, "Virginity Lost," *Conservation* 8, no. 1 (January–March 2007), http:// www.conservationmagazine.org/articles/v8n1/virginity-lost/ (accessed February 19, 2009).

6. Pearce, "Virginity Lost," *Conservation* 8, no 1 (January–March 2007), http:// www.conservationmagazine.org/articles/v8n1/virginity-lost/ (accessed February 19, 2009).

7. Stephen Leahy, "Climate Change: Will Forests Adapt to a Warmer World?" Inter Press Service, November 20, 2006, http://www.ipsnews.net/news.asp?idnews =35548 (accessed July 23, 2009).

8. Rhett A. Butler, "Deforestation a Greater Threat to the Amazon than Global Warming," mongabay.com, February 25, 2008, http://news.mongabay.com/2008/ 0225-mayle_amazon.html (accessed July 23, 2009).

9. Ibid.

10. "Global Warming May Drive the Amazon Rainforest toward Seasonal Forests Rather than Savanna," mongabay.com, February 11, 2009, http://news.mongabay .com/2009/0211-amazon.html (accessed July 23, 2009).

11. Richard Alleyne, "Amazon Rainforest at Risk of Ecological 'Catastrophe,'" *Telegraph*, March 12, 2009, http://www.telegraph.co.uk/earth/environment/ climatechange/4976275/Amazon-rainforest-at-risk-of—ecological- catastrophe.html (accessed July 23, 2009).

12. "Scientists to Set Fire to Amazon Rainforest to Study Its Resilience," press release from the Woods Hole Research Center, Mongabay, July 19, 2005, http:// news.mongabay.com/2005/0719-whrc.html (accessed July 23, 2009); Stephen Leahy, "Climate Change: Will Forests Adapt to a Warmer World?" Inter Press Service, Nov 20, 2006, http://www.ipsnews.net/news.asp?idnews=35548 (accessed July 23, 2009).

13. "Climate Change Impacts in the Amazon: Review of scientific literature," World Wildlife Fund website, http://assets.panda.org/downloads/amazon_cc_impacts _lit_review_final_2.pdf (accessed January 8, 2009).

14. "Amazon River in Peru Falls to Second Lowest Level on Record," mongabay.com, September 4, 2007, http://news.mongabay.com/2007/0904-amazon.html (accessed April 15, 2008).

15. David Fogarty, "Emerging El Niño Set to Drive Up Carbon Emissions," Reuters, July 7, 2009, http://in.reuters.com/article/environmentNews/idINTRE5660432 0090707 (accessed July 23, 2009).

16. Pew Institute For Ocean Science, Directory of Fellows Philip Fearnside, http:// www.pewoceanscience.org/fellows/pfearnside/fellows-dir-profile.php?pfID =3559 (accessed December 9, 2008).

17. Interview with Philip Fearnside, Manaus, October 24, 2008.
18. Fogarty, "Emerging El Niño," Reuters, July 7, 2009, http://in.reuters.com/article/environmentNews/idINTRE56604320090707 (accessed July 23, 2009).
19. "Indonesia: Drought—OCHA–05: 06-May–98, INDONESIA EL NIÑO—Drought in Irian Jaya," OCHA-Geneva Situation Report No. 5, May 6, 1998, Center for International Disaster Information website, http://cidi.org/disaster/98a/0071.html (accessed January 16, 2009).
20. Joelle Diderich, "Yanomami's Turn to Shamans to Stop Amazon Fires," Reuters, March 19, 1998, http://forests.org/archive/brazil/turnsham.htm (accessed January 12, 2009).
21. "Rainforest Fire Threatens Amazon Indians," BBC, March 5, 1998, http://news.bbc.co.uk/2/hi/despatches/62571.stm (accessed January 12, 2009); "El Niño sigue provocando inundaciones en Perú, El Ejército brasileño quiere impedir el acceso de extranjeros al incendio de la Amazonía," *El País*, January 27, 1998, http://www.elpais.com/articulo/sociedad/PERu/LATINOAMERICA/BOLIVIA/ECUADOR/Nino/sigue/provocando/inundaciones/Peru/elpepisoc/19980327elpepisoc_3/Tes/ (accessed January 7, 2009).
22. Dan Nepstad, "Climate Change and the Forest," *The American Prospect*, September 1, 2007, http://www.allbusiness.com/environment-natural-resources/ecology/5523170–1.html (accessed January 12, 2009).
23. Diderich, "Yanomami's Turn to Shamans," Reuters, March 19, 1998, http://forests.org/archive/brazil/turnsham.htm (accessed January 12, 2009).
24. Nepstad, "Climate Change and the Forest," *American Prospect*, September 1, 2007, http://www.allbusiness.com/environment-natural-resources/ecology/5523170–1.html (accessed January 12, 2009).
25. Diderich, "Yanomami's Turn to Shamans," Reuters, March 19, 1998, http://forests.org/archive/brazil/turnsham.htm (accessed January 12, 2009).
26. "Brazil–Roraima Fires and Drought," CWS/ACT APPEAL: BRAZIL–RORAIMA FIRES AND DROUGHT (#976309), ReliefWeb website, May 22, 1998, http://www.reliefweb.int/rw/rwb.nsf/db900sid/OCHA64CH4K?OpenDocument (accessed January 12, 2009).
27. "Guyana: El Niño Drought," appeal no: 14/98, International Federation of Red Cross and Red Crescent Societies website, April 22, 1998, http://www.ifrc.org/docs/appeals/98/1498.pdf (accessed January 15, 2009).
28. "Biology: Studies in the Area of Biology Reported from J.A. Marengo and Co-researchers," *Life Science Weekly*, May 13, 2008; Larry Rohter, "A Record Amazon Drought, and Fear of Wider Ills," *New York Times*, December 11, 2005.
29. Peter M. Cox, Phil P. Harris, Chris Huntingford, Richard A. Betts, Matthew Collins, Chris D. Jones, Tim E. Jupp, José A. Marengo, and Carlos A. Nobre, "Increasing Risk of Amazonian Drought Due to Decreasing Aerosol Pollution," *Nature* 453 (May 8, 2008): 212–215, http://www.nature.com/nature/journal/v453/n7192/full/nature06960.html (accessed January 8, 2009).
30. Brown, Schroeder, Setzer, "Monitoring Fires in Southwestern Amazonia Rain Forests," *EOS, Transactions, American Geophysical Union* 87, no. 26 (June 27, 2006): 253–264, Centro de Previsão de Tempo e Estudos Climáticos–CPTEC/INPE website, http://www.cptec.inpe.br/queimadas/documentos/200606_brown_schroeder_setzer_outros_eos_87_26.pdf (accessed January 4, 2009).
31. Alex Shoumatoff, "The Gasping Forest," *Vanity Fair*, May 2007, http://www.vanityfair.com/politics/features/2007/05/amazon200705 (accessed January 4, 2009).

32. Interview with Fernando Rodríguez, Iquitos, October 17, 2008.
33. Brown, Schroeder, Setzer, "Monitoring Fires in Southwestern Amazonia Rain Forests," *EOS, Transactions, American Geophysical Union* 87, no. 26 (June 27, 2006): 253–264, Centro de Previsão de Tempo e Estudos Climáticos–CPTEC/ INPE website, http://www.cptec.inpe.br/queimadas/documentos/200606_brown _schroeder_setzer_outros_eos_87_26.pdf (accessed January 4, 2009).
34. "Amazon River in Peru Falls to Second Lowest Level on Record," mongabay.com, September 4, 2007, http://news.mongabay.com/2007/0904-amazon.html (accessed April 8, 2008); Rohter, "A Record Amazon Drought," *New York Times,* December 11, 2005.
35. Zoraida Portillo, "Drought Threatens Amazon Carbon Sink," Sci Dev website, March 9, 2009, http://www.scidev.net/en/news/drought-threatens-amazon-carbon-sink.html (accessed July 23, 2009).
36. "Amazon Rainforest Carbon Sink Threatened By Drought," Science Daily, March 9, 2009, http://www.sciencedaily.com/releases/2009/03/090305141625 .htm (accessed July 23, 2009).
37. Interview with Janet Larsen, Washington, D.C., August 26, 2008.
38. Rhett Butler, "Study Discovers Why Poison Dart Frogs Are Toxic," mongabay .com, August 9, 2005, http://news.mongabay.com/2005/0809-frogs (accessed January 8, 2009).
39. Interview with Larsen, Washington, D.C., August 26, 2008.
40. "Hot, Hungry and Gasping for Air—Climate Change Puts Fish at Risk, Warns WWF," World Wildlife website, November 18, 2005, http://www.panda.org/ about_wwf/what_we_do/climate_change/?uNewsID=50460 (accessed January 14, 2009).
41. Interview with Juan Berchota, Iquitos, October 17, 2008.
42. "In Peru, a Move to Get Farmers to Trade in Fish Rather than Coca," *Christian Science Monitor,* http://www.csmonitor.com/2007/0104/p05s01-woam.html (accessed January 4, 2009).
43. "Villagers and Wildlife at Risk," *Times* (London), October 12, 2005, http://www .timesonline.co.uk/tol/news/world/us_and_americas/article577534.ece (accessed January 4, 2009).
44. Jeremy Hance, "An Interview with Dr. Jay Barlow: Extinction of the Baiji a 'Wake-Up Call' to Conserve Vaquita and Other Cetaceans," mongabay.com, August 25, 2008 (accessed January 19, 2009).
45. Andrew C. Revkin, "The Climate Divide, Wealth and Poverty, Drought and Flood: Reports From 4 Fronts in the War on Warming," *New York Times,* April 3, 2007, http://query.nytimes.com/gst/fullpage.html?res=9C03E1DE1F30F93 0A35757C0A9619C8B63 (accessed January 14, 2009).
46. "Tropical Deforestation Affects Rainfall in the U.S. and around the Globe," Science Daily, September 18, 2005, http://www.sciencedaily.com/releases/2005/ 09/050918132252.htm (accessed June 28, 2009).
47. Indira A.R. Lakshmanan, "Amazon Burning," *Boston Globe,* November 27, 2006, http://www.boston.com/news/globe/health_science/articles/2006/11/27/ amazon_burning/ (accessed June 28, 2009); Axel Bugge, "Amazon Fires Alter Landscape in Many Ways," Reuters, July 28, 2004, http://www.abc.net.au/ science/news/stories/s1163886.htm (accessed June 28, 2009); "Tropical Deforestation Affects Rainfall in the U.S. and around the Globe," Science Daily, September 18, 2005, http://www.sciencedaily.com/releases/2005/09/050918132252 .htm (accessed June 28, 2009).

CHAPTER 4

1. Rhett A. Butler, "Amazon Conservation Efforts Must Come Soon to Save World's Largest Rainforest Says Leading Scientist," mongabay.com, October 23, 2006, http://news.mongabay.com/2006/1023-interview_fearnside.html (accessed March 20, 2009).

2. Greenpeace, "Slaughtering the Amazon," part one, updated July 2009 version, http://www.greenpeace.org/international/assets/binaries/slaughtering-the-amazon-part1 (accessed July 24, 2009).

3. Jim Motavalli, "The Meat of the Matter," *Orlando Weekly,* July 31, 2008, http://www.orlandoweekly.com/features/story.asp?id=12523 (accessed March 20, 2009).

4. "Brazil Declares Forest Havens after Nun's Killing," *The Guardian,* February 19, 2005, http://www.guardian.co.uk/environment/2005/feb/19/endangeredhabitats.activists (accessed March 20, 2009).

5. Matt Moffett, "Brazil's Grass-fed Cattle Are Its Economic Salvation—Beef Exports Are Soaring," *The Wall Street Journal,* June 22, 2004, posted at mongabay .com, http://www.mongabay.com/external/brazils_soaring_beef_exports.htm (accessed March 20, 2009).

6. "World Bank Close to Approving Amazon Beef, Other Projects," Dow Jones Commodities Service, January 22, 2007.

7. "Amazon Beef Exports Get OK," Dow Jones Commodities Service, May 22, 2007.

8. "World Bank Pledges to Save Trees . . . Then Helps Cut Down Amazon Forest," Environmental News Network, January 16, 2008, http://www.enn.com/ecosystems/article/29402 (accessed March 20, 2009).

9. Greenpeace, "Slaughtering the Amazon," part one, updated July 2009 version, http://www.greenpeace.org/international/assets/binaries/slaughtering-the-amazon-part1 (accessed July 24, 2009).

10. Greenpeace, "Slaughtering the Amazon," part one, updated July 2009 version, http://www.greenpeace.org/international/assets/binaries/slaughtering-the-amazon-part1 (accessed July 24, 2009).

11. Raymond Colitt, "Brazil Seizes Cattle to Stem Amazon Destruction," Reuters, June 24, 2008, http://www.reuters.com/article/environmentNews/idUSN24363 93820080624 (accessed August 1, 2009).

12. Nick Caistor, "Brazil's 'slave' ranch workers," BBC, May 11, 2005, http://news.bbc.co.uk/2/hi/americas/4536085.stm (accessed March 20, 2009).

13. Ibid.

14. Danielle Calentano and Adalberto Veríssimo, "The Amazon Frontier Advance: From Boom to Bust," Imazon, Amazon Institute of People and the Environment, http://www.amazonia.org.br/arquivos/258120.pdf (accessed March 20, 2009).

15. Tales Azzoni, "American Missionary from Ohio Killed in Northern Brazil," Associated Press Newswires, February 12, 2005; Jan Rocha, "Obituary: Sister Dorothy Stang," *The Guardian,* February 21, 2005, http://www.amazonia.org.br/english/noticias/noticia.cfm?id=147106 (accessed March 20, 2009); Sherri Williams, "Ohio Family Proudly Recalls Spirit of Nun Killed in Brazil," *Columbus Dispatch,* February 16, 2005; Paulo Prado, "Murdered Nun Fought for Rights, Environment," *Miami Herald,* February 15, 2005; Andrew Buncombe, "The Life and Brutal Death of Sister Dorothy, a Rainforest Martyr," *The Independent,* February 15, 2005, http://www.amazonia.org.br/english/noticias/noticia.cfm?id=145694 (accessed March 20, 2009).

16. Michael Astor, "Brazil Races to Investigate Murder of American Missionary," Associated Press, February 13, 2005.
17. "Rainforest Martyr," web site devoted to documentary film *The Student, the Nun and the Amazon* about Dorothy Stang, http://www.studentnunamazon.com/data/pages/martyr.htm (accessed March 20, 2009).
18. Buncombe, "The Life and Brutal Death of Sister Dorothy," *Independent,* February 15, 2005, amazonia.org, http://www.amazonia.org.br/english/noticias/noticia.cfm?id=145694 (accessed March 20, 2009).
19. Jan Rocha, "Brazil Declares Forest Havens after Nun's Killing," *Guardian,* February 19, 2005, http://www.guardian.co.uk/environment/2005/feb/19/endangeredhabitats.activists (accessed March 20, 2009).
20. Prada, "Murdered nun fought for rights," *The Miami Herald,* February 15, 2005, amazonia.org, http://www.amazonia.org.br/english/noticias/noticia.cfm?id=145684 (accessed March 20, 2009).
21. "Rainforest Martyr," web site devoted to documentary film *The Student, the Nun and the Amazon* about Dorothy Stang, http://www.studentnunamazon.com/data/pages/martyr.htm (accessed March 20, 2009).
22. Prada, "Murdered Nun Fought for Rights," *Miami Herald,* February 15, 2005, amazonia.org, http://www.amazonia.org.br/english/noticias/noticia.cfm?id=145684 (accessed March 20, 2009).
23. "Sister Dorothy Stang, Missionary and Defender of the Amazon Rainforest in Brazil," *Independent,* February 15, 2005, http://www.amazonia.org.br/english/noticias/noticia.cfm?id=145695 (accessed March 20, 2009).
24. "Rainforest Martyr," web site devoted to documentary film *The Student, the Nun and the Amazon* about Dorothy Stang, http://www.studentnunamazon.com/data/pages/martyr.htm (accessed March 20, 2009).
25. Rocha, "Brazil Declares Forest Havens after Nun's Killing," *The Guardian,* February 19, 2005, http://www.guardian.co.uk/environment/2005/feb/19/endangeredhabitats.activists (accessed March 20, 2009).
26. Tom Phillips, "Brazil's Burden . . . Slavery," *Sunday Herald,* undated, http://www.sundayherald.com/international/shinternational/display.var.2490875.0.0.php (accessed March 20, 2009).
27. "Cowboys and Land-grab," BBC, http://www.bbc.co.uk/amazon/sites/cowboys/pages/content.shtml (accessed March 20, 2009).
28. Interview with Alex Webb, Brooklyn, NY, September 10, 2008.
29. "Cowboys and Land-grab," BBC, http://www.bbc.co.uk/amazon/sites/cowboys/pages/content.shtml (accessed March 20, 2009).
30. Diana Schemo, "Marina Silva," *Ms.* 56, 8, no. 4 (January 1, 1998).
31. Ibid.
32. Michael Astor, "In Brazil, a Daughter of the Jungle Becomes Its Most Powerful Guardian," Associated Press, January 25, 2003.
33. "Enviros Force Brazilian Congress to Kill Destructive Rainforest Law," Environment News Service, May 18, 2000.
34. Raymond Colitt, "World's Largest Wetland Threatened in Brazil," Reuters, February 16, 2009, http://uk.news.yahoo.com/22/20090216/twl-environment-us-brazil-pantanal–1202b49.html (accessed March 20, 2009).
35. "Scientists Warn of Wetlands 'Carbon Bomb,'" *Telegraph,* July 20, 2008, http://www.telegraph.co.uk/earth/environment/3347681/Scientists-warn-of-wetlands-carbon-bomb.html (accessed March 20, 2009).
36. Ibid.

37. "Pantanal May Become Next Everglades, UNU Experts Warn," Update.unu.edu (newsletter of United Nations University and its international network of research and training centres/programmes), no. 37, May–June 2005, http://update .unu.edu/default.htm (accessed March 20, 2009); Stephen Leahy, "Warming Threatens Precious Wetlands," Inter Press Service, March 22, 2005, http:// www.ipsnews.net/interna.asp?idnews=27972 (accessed March 20, 2009).

38. "Farming Destroying Brazil's Wetlands," United Press International, January 11, 2006, posted at phys.org website, http://www.physorg.com/news9773.html (accessed March 20, 2009).

39. "Success Stories: The Hyacinth Macaw Makes a Comeback," World Wildlife Fund, April 21, 2004, http://www.panda.org/what_we_do/successes/?12641 (accessed March 20, 2009); Jeremy Hance, "Predator of the World's Largest Macaw Key to Its Survival," mongabay.com, March 13, 2008, http://news.mongabay .com/2008/0313-hance_macaws.html (accessed August 1, 2009).

40. Raymond Colitt, "Amazon Defender Quits Brazil Environment Post," Reuters, May 13, 2008, http://www.reuters.com/article/latestCrisis/idUSN13420813 (accessed March 20, 2009).

41. Andrew Downie, "Brazil Seizes Livestock to Protect Rain Forest," *New York Times,* June 25, 2008, http://www.nytimes.com/2008/06/25/business/world-business/25beef.html (accessed March 20, 2009).

42. Ibid.

43. Greenpeace, "Slaughtering the Amazon," part one, updated July 2009 version, http://www.greenpeace.org/international/assets/binaries/slaughtering-the-amazon-part1 (accessed July 24, 2009).

44. Ibid.

45. Ramesh Jaura, "Climate Change: 'Don't Leave it to the World Bank,'" Inter Press Service, December 10, 2008, http://ipsnews.net/news.asp?idnews=45050 (accessed July 28, 2009).

46. "Nike, Unilever, Burger King, IKEA may unwittingly contribute to Amazon destruction, says Greenpeace," mongabay.com, June 1, 2009, http://news.mongabay .com/2009/0601-greenpeace_beef.html (accessed July 24, 2009).

47. Greenpeace, "Slaughtering the Amazon," part one, updated July 2009 version, http://www.greenpeace.org/international/assets/binaries/slaughtering-the-amazon-part1 (accessed July 24, 2009).

48. Ibid.

49. Ibid.

50. Ibid.

51. Ibid.

52. Ibid.

53. Ibid.

54. Ibid.

55. "Nike, Unilever, Burger King, IKEA may unwittingly contribute to Amazon destruction, says Greenpeace," mongabay.com, June 1, 2009, http://news.mongabay .com/2009/0601-greenpeace_beef.html (accessed July 24, 2009).

CHAPTER 5

1. Pat Joseph, "Soy in the Amazon," *Virginia Quarterly Review,* fall 2007, http:// www.vqronline.org/articles/2007/fall/joseph-soy-amazon/ (accessed April 11, 2009); Forest Forest Footprint Disclosure Project (UK government-supported

initiative, created to help investors identify how an organization's activities and supply chains contribute to deforestation) website, http://forestdisclosure.com/soy.html (accessed October 5, 2009).

2. "Growing Trade Ties China to Latin America," National Public Radio, *All Things Considered*, April 1, 2008.

3. Joseph, "Soy in the Amazon," *Virginia Quarterly Review*, fall 2007, http://www.vqronline.org/articles/2007/fall/joseph-soy-amazon/ (accessed April 12, 2009); "Eating Up the Amazon," Greenpeace, http://www.greenpeace.org/raw/content/international/press/reports/eating-up-the-amazon.pdf (accessed April 12, 2009).

4. "King of Soya: Environmental Vandal or Saviour?" *Mail and Guardian*, March 3, 2008.

5. David Munk, "Society: Environment: Forces of nature: Brazil's Rich Environment Is under Attack from All Sides. David Munk Meets Marina Silva, the Minister Who Must Restore the Balance," *Guardian*, March 17, 2004.

6. "Prize Recognizes Largest Contributor to Amazon Rainforest Destruction," mongabay.com, May 27, 2005, http://news.mongabay.com/2005/0527-golden_chainsaw.html (accessed April 12, 2009).

7. "IFC Green Gloss: New Private Sector Guide to Biodiversity," Bretton Woods Project, March 26, 2006, http://www.brettonwoodsproject.org/art.shtml?x=531506 (accessed August 4, 2009).

8. "IFC Soy Project subject of Controversy—Categorization as class B under Fire," Bank Information Center, March 22, 2005, http://www.bicusa.org/en/Article.1990.aspx (accessed August 4, 2009).

9. "Prize Recognizes Largest Contributor to Amazon Rainforest Destruction," mongabay.com, May 27, 2005, http://news.mongabay.com/2005/0527-golden_chainsaw.html (accessed April 12, 2009).

10. "Eating Up the Amazon," Greenpeace, http://www.greenpeace.org/raw/content/international/press/reports/eating-up-the-amazon.pdf (accessed April 12, 2009).

11. Carlos Coimbra, Nancy Flowers, and Francisco Salzano, *The Xavante in Transition: Health, Ecology and Bioanthropology in Central Brazil* (Ann Arbor: University of Michigan Press, 2004), Google Books, p. 72, http://books.google.com/books?id=PSEJEX2NGlgC&pg=PA72&dq=peter+fleming+mato+grosso (accessed April 12, 2009).

12. Alex Bellos, "The Long Road to Ruin," *Observer*, April 15, 2007, http://www.guardian.co.uk/environment/2007/apr/15/brazil.conservation (accessed April 12, 2009).

13. Gonçalves, "Driving along Brazil's Highway BR–163," World Wildlife Fund, May 15, 2007, http://www.panda.org/wwf_news/features/?100120/Driving-along-Brazils-Highway-BR–163 (accessed August 4, 2009).

14. Alan Clendenning, "Road Cuts Deep into Brazil's Amazon: Settlers Expect Wealth, Environmentalists Disaster from Paving," Associated Press, May 28, 2005, posted at mongabay. com, http://news.mongabay.com/2005/0527-ap.html (accessed April 12, 2009).

15. Scott Wallace, "Farming the Amazon," *National Geographic*, January 2007, http://environment.nationalgeographic.com/environment/habitats/last-of-amazon.html (accessed April 12, 2009).

16. Indira Lakshmanan, "Amazon Highway Is Route to Strife in Brazil," *Boston Globe*, December 27, 2005, http://www.boston.com/news/world/articles/2005/12/27/amazon_highway_is_route_to_strife_in_brazil/ (accessed April 12, 2009).

17. Alex Bellos, "Blood Crop: Violent Side of Brazil's Soya Industry," *Telegraph,* October 12, 2007, http://www.telegraph.co.uk/earth/3310375/Blood-crop-Violent-side-of-Brazils-soya-industry.html (accessed April 12, 2009).

18. "Greenpeace Activists in Brazil Block Cargill Soy Facility," Greenpeace, May 22, 2006, http://www.greenpeace.org/international/news/greenpeace-activists-in-brazil (accessed April 12, 2009).

19. Michael Grunwald, "The Clean Energy Scam," *Time,* March 27, 2008, http://www.time.com/time/magazine/article/0,9171,1725975–2,00.html (accessed April 12, 2009).

20. "Taking on Tofu: Can Soy Be Sustainable?" *Canopy* (Quarterly Publication of the Rainforest Alliance). 17, no. 3 (fall 2004), http://www.rainforest-alliance.org/soy.html (accessed April 12, 2009).

21. Joshua Schneyer, "Brazil's Answer to Global Hunger," *Business Week* 72, vol. 4086, June 2, 2008.

22. "Landmark Amazon Soy Moratorium Extended," Greenpeace, June 17, 2008, http://www.greenpeace.org/usa/news/landmark-amazon-soya-moratoriu (accessed April 12, 2009).

23. Susanna Hecht and Charles Mann, "How Brazil Outfarmed the American Farmer," *Fortune,* January 19, 2008, http://money.cnn.com/2008/01/16/news/international/brazil_soy.fortune/index.htm (accessed April 12, 2009).

24. Jan Maarten Dros, "Managing the Soy Boom: Two Scenarios of Soy Production Expansion in South America," World Wildlife Fund, June 2004, http://assets.panda.org/downloads/managingthesoyboomenglish_nbvt.pdf (accessed April 12, 2009).

25. Alex Morales and Jeremy van Loon, "Rain Forest–Saving Credits May Cut Carbon-Emission Prices," Bloomberg, March 31, 2009, http://www.bloomberg.com/apps/news?pid=20601072&sid=aYhjvwNIwjUo&refer=energy (accessed August 4, 2009).

26. Alan Clendenning, "Brazil President Defends Biofuels at Climate Meeting, Says Production Won't Hurt Amazon," Associated Press Newswires, February 21, 2008.

27. "Brazil Not to Plant Sugar Cane in Amazon: President," Xinhua News Agency, June 2, 2008.

28. Rhett Butler, "Big REDD," *Washington Monthly,* July/August 2009, http://www.washingtonmonthly.com/features/2009/0907.butler.html (accessed October 6, 2009)

29. Ibid.

30. Ibid.

31. Roger Harrabin, "Anger over Climate Change Loans," BBC, http://news.bbc.co.uk/2/hi/uk_news/politics/7407528.stm (accessed August 4, 2009).

32. "All About: Forests and Carbon Trading," CNN, February 11, 2008, http://www.cnn.com/2008/TECH/02/10/eco.carbon/ (accessed May 27, 2009).

33. Ramesh Jaura, "Climate Change: 'Don't Leave it to the World Bank,'" Inter Press Service, December 10, 2008, http://ipsnews.net/news.asp?idnews=45050 (accessed July 28, 2009).

34. Butler, "Big REDD," *Washington Monthly,* July/August 2009, http://www.washingtonmonthly.com/features/2009/0907.butler.html.

CHAPTER 6

1. "The Amazon in Danger," *New York Times,* November 20, 2006, http://travel.nytimes.com/frommers/travel/guides/central-and-south-america/peru/

amazon-basin/frm_amazon-bas_2882010012.html?pagewanted=print (accessed July 7, 2009); James Painter, "Peru Aims for Zero Deforestation," BBC, http://news.bbc.co.uk/2/hi/americas/7768226.stm (accessed June 27, 2009).

2. Max Milligan, "Inca Spots," *Guardian,* September 29, 2001, http://www.guardian.co.uk/travel/2001/sep/29/peru.wildlifeholidays.guardiansaturdaytravelsection (accessed June 27, 2009).

3. Paula Dobriansky, "Earth Day Equation: Drug Abuse = Environmental Abuse," *Christian Science Monitor,* April 21, 2003, http://www.csmonitor.com/2003/0421/p11s01-coop.html (accessed June 27, 2009).

4. "Clarity Brought to Earth's Cloud Forests," United Nations Environment Program, February 9, 2004, http://new.unep.org/Documents.Multilingual/Default_pub.asp?DocumentID=384&ArticleID=4406&l=en (accessed June 27, 2009).

5. "Illegal Drug Use Destroys Rainforests," mongabay.com, November 18, 2008, http://news.mongabay.com/2008/1118-cocaine.html (accessed June 27, 2009).

6. Andrew Whalen, "Peru's Highland Rebels Feed Off Cocaine Trade," Associated Press, May 23, 2009, http://www.timesdispatch.com/rtd/lifestyles/health_med_fit/article/I-PERU0508_20090521-205813/269147/ (accessed June 27, 2009).

7. Jaime Cordero, "El 'narco' destruye dos millones de hectáreas en Perú," *El País,* December 31, 2008, http://www.elpais.com/articulo/internacional/narco/destruye/millones/hectareas/Peru/elpepuint/20081231elpepuint_3/Tes (accessed June 27, 2009).

8. Juan C. De La Cal, "La caoba sale cara 56.000 esclavos, mujeres que se prostituyen por restos de madera, la selva asolada . . . y España, a la cabeza del consumo," *El Mundo,* August 19, 2007, http://www.elmundo.es/suplementos/cronica/2007/617/1187474406.html (accessed June 27, 2009); E.U. Curtis Bohlen and David O. Sandalow, "Protecting Mahogany: Bush Policy Sells Amazon Treasure Down the River," *New York Times,* November 13, 2002, http://www.nytimes.com/2002/11/13/opinion/13iht-edcurtis_ed3_.html (accessed June 27, 2009).

9. "Selective Logging Causes Widespread Destruction of Brazil's Amazon Rainforest, Study Finds," NASA, October 20, 2005, http://earthobservatory.nasa.gov/Newsroom/view.php?id=28547 (accessed June 27, 2009).

10. "Don't Forget the Second 'D,'" Nature Conservancy Policy Brief, June 2009, http://www.nature.org/initiatives/protectedareas/files/tnc_degradation_policy_brief_lowres.pdf (accessed June 27, 2009).

11. Juan C. De La Cal, "La caoba sale cara 56.000 esclavos," *El Mundo,* August 19, 2007, http://www.elmundo.es/suplementos/cronica/2007/617/1187474406.html (accessed June 27, 2009).

12. "Peru: Illegal logging—a source of forced labour in the Amazon," World Rainforest Movement, http://www.wrm.org.uy/bulletin/99/Peru.html (accessed June 28, 2009).

13. Milligan, "Inca Spots," *Guardian,* September 29, 2001, http://www.guardian.co.uk/travel/2001/sep/29/peru.wildlifeholidays.guardiansaturdaytravelsection (accessed June 27, 2009).

14. Andrés Schipani and John Vidal, "Malaria Moves in Behind the Loggers," *Guardian,* October 30, 2007, http://www.guardian.co.uk/world/2007/oct/30/environment.climatechange (accessed June 28, 2009).

15. Bohlen and Sandalow, "Protecting Mahogany: Bush Policy Sells Amazon Treasure Down the River," *New York Times,* November 13, 2002, http://www.nytimes.com/2002/11/13/opinion/13iht-edcurtis_ed3_.html (accessed June 27, 2009).

16. Juan Forero, "A Swirl of Foreboding in Mahogany's Grain," *New York Times*, September 28, 2003, http://www.nytimes.com/2003/09/28/world/a-swirl-of-foreboding-in-mahogany-s-grain.html (accessed June 27, 2009).

17. Rhett A. Butler, "Peru Fails to Investigate Murder of Amazon Environmental Leader," mongabay.com, April 22, 2008, http://news.mongabay.com/2008/0422-julio_garcia.html (accessed June 28, 2009); "Our Role in Stopping the Illegal Brazilian Mahogany Trade," Greenpeace website, October 15, 2003, http://www.greenpeace.org/usa/news/our-role-in-stopping-the-illeg (accessed October 8, 2009).

18. James Painter, "Peru Aims for Zero Deforestation," BBC, December 7, 2008, http://news.bbc.co.uk/2/hi/americas/7768226.stm (accessed June 28, 2009).

19. Interview with Laura Avellaneda, Lima, Peru, October 1, 2008.

20. "Perfil de Antonio Brack," University of Lima website, http://fresno.ulima.edu.pe/sf/sf_bd5200.nsf/OtrosWeb/np/$file/np.pdf (accessed December 15, 2008).

21. Butler, "Peru to Raise Payment to Indigenous Communities for Amazon Forest Conservation," mongabay.com, August 3, 2009, http://news.mongabay.com/2009/0803-peru.html (accessed August 7, 2009).

22. Painter, "Peru Aims for Zero Deforestation," BBC, http://news.bbc.co.uk/2/hi/americas/7768226.stm (accessed June 28, 2009).

23. "Tribes in Peru to Get $0.68/acre for Protecting Amazon Forest," mongabay.com, June 3, 2009, http://news.mongabay.com/2009/0603-peru.html (accessed June 28, 2009).

24. "Camisea Fuels Oil Boom in Peru," IDB Watch, Issue Two April 5, 2008, Amazon Watch, http://www.amazonwatch.org/documents/IDBWatch_Issue2.pdf (accessed August 7, 2009).

25. "Description of Camisea project," Friends of the Earth, http://www.foe.org/camps/intl/institutions/camisea.htm (accessed August 7, 2009).

26. "Letter of Demands to Enrique Iglesias," Bank Information Center, December 3, 2003, http://www.bicusa.org/en/Article.365.aspx (accessed August 7, 2009).

27. "Camisea Natural Gas Project," Amazon Watch website, http://www.amazonwatch.org/amazon/PE/camisea/ (accessed August 7, 2009)

28. Lucien Chauvin, "Peru Natural Gas Pipelines, Plant Ignite Controversy," *San Francisco Chronicle*, August 5, 2003, http://www.sfgate.com/cgi-bin/article.cgi?f=/c/a/2003/08/05/MN260268.DTL (accessed August 7, 2009).

29. "Unaccountable and Unsustainable, IDB Champions Corporate Interests at Expense of Citizens," Bank Information Center, April 3, 2008, http://www.bicusa.org/en/Article.3715.aspx (accessed August 7, 2009).

30. Interview with Andrew Miller of Amazon Watch, Washington, D.C., August 26, 2008.

31. Interview with Carlos Alza, Lima, Peru, September 29, 2008.

32. "Peru Tribe Battles Oil Giant Over Pollution," BBC, March 24, 2008, http://news.bbc.co.uk/2/hi/americas/7306639.stm (accessed December 23, 2008).

33. "Amazonian Tribe Protests at Oil Pollution," BBC, September 13, 2006, http://news.bbc.co.uk/2/hi/americas/5337802.stm (accessed December 23, 2008).

34. Milagros Salazar, "Peru: 'Police Are Throwing Bodies in the River,'" Tierramérica, June 9, 2009, http://upsidedownworld.org/main/content/view/1901/1/ (accessed June 28, 2009).

35. Amazon Watch, AIDESEP, "Eyewitness Reports Accuse Peruvian Police of Disposing the Bodies of Dead Indigenous Protesters," Amazon Watch, June 8, 2009,

http://www.amazonwatch.org/newsroom/view_news.php?id=1843 (accessed June 28, 2009).

36. "Rainforest Resources Conflict in Northern Peru Turns Bloody," Environmental News Services, June 6, 2009, http://www.ens-newswire.com/ens/jun2009/2009–06–06–01.asp (accessed June 28, 2009).

37. "Peru Approves Controversial Amazon Oil Contract—in Wake of Uprising," World War 4 Report, July 3, 2009, http://www.ww4report.com/node/7533 (accessed July 7, 2009).

CHAPTER 7

1. Larry Rohter, "Brazil, Fearful of Blackouts, Orders 20 Percent Cut in Electricity," New York Times, May 19, 2001, http://www.nytimes.com/2001/05/19/world/brazil-fearful-of-blackouts-orders–20-cut-in-electricity.html?n=Top percent2FReference percent2FTimes percent20Topics percent2FSubjects percent2FF percent2FFinances (accessed April 30, 2009).

2. Marcelo Ballvé, "Mega-Dams in the Amazon: Brazil's Three Gorges?" NACLA, July 26 2007, http://nacla.org/node/1501 (accessed February 14, 2009).

3. "The Enawene Nawe," Survival International, http://www.survival-international.org/tribes/enawenenawe (accessed April 26, 2009); Christina Lamb, "The tribe that stood their ground," The Sunday Times, February 15, 2009, http://www.timesonline.co.uk/tol/news/world/us_and_americas/article5716227.ece (accessed April 26, 2009).

4. Sanjeev Khagram, Dams and Development (Ithaca: Cornell University Press, 2004), 155, Google Books, http://books.google.com/books?id=TPSCkBYFE-SUC&pg=PA155&lpg=PA155&dq=waimiri+atroari+balbina&source=bl&ots=MhueOI76vQ&sig=vlHoiUhdzLgbaSk4GZ5cdvTdqNU&hl=en&ei=sd_kS-fyNEJDglQfsgd3gDg&sa=X&oi=book_result&ct=result&resnum=9 (accessed April 26, 2009).

5. Philip M. Fearnside, "Social and Environmental Impacts of Hydroelectric Dams in Brazilian Amazonia," Inpa website, http://philip.inpa.gov.br/publ_livres/1999/hydro percent20German-english.pdf (accessed February 14, 2009).

6. "The Amazon: The Facts," New Internationalist, no. 219 (May 1991), http://www.newint.org/issue219/facts.htm (accessed February 14, 2009); Marlise Simons, "Dam's Threat to Rain Forest Spurs Quarrels in the Amazon," New York Times, September 6, 1987, http://query.nytimes.com/gst/fullpage.html?res=9B0DE1D61139F935A3575AC0A961948260 (accessed February 14, 2009).

7. Philip Fearnside, "Profile: Brazil's Balbina Dam: Environment versus the Legacy of the Pharaohs in Amazonia," Environmental Management 13, no. 4 (July 1989), http://www.springerlink.com/content/l0504508pv1632u2/fulltext.pdf?page=1 (accessed February 14, 2009).

8. Barbara Fraser, "Glaciers Go, Leaving Drought, Conflict and Tension in Andes," The Daily Climate, May 19, 2009, http://wwwp.dailyclimate.org/tdc-newsroom/2009/05/glaciers-go-leaving-drought-conflict-and-tension (accessed August 9, 2009).

9. Fearnside, "Profile: Brazil's Balbina Dam: Environment versus the Legacy of the Pharaohs in Amazonia," Environmental Management 13, no. 4 (July 1989), http://www.springerlink.com/content/l0504508pv1632u2/fulltext.pdf?page=1 (accessed February 14, 2009).

10. Bruce Barcott, "Opinion: Big, Bad Hydro," *Forbes,* June 19, 2008, http://www .forbes.com/2008/06/19/dams-methane-warming-tech-water08-cx_bb_0619 dam.html (accessed April 26, 2009).

11. Reese Ewing, "Brazil to Flood Amazon Rainforest for Hydroelectric Power," Reuters, March 17, 2006, http://news.mongabay.com/2006/0317-reuters.html (accessed February 14, 2008).

12. Barcott, "Opinion: Big, Bad Hydro," *Forbes,* June 19, 2008, http://www.forbes .com/2008/06/19/dams-methane-warming-tech-water08-cx_bb_0619dam.html (accessed April 26, 2009).

13. "Brazil: IIRSA Madeira River Complex," Amazon Watch web site, http://www .amazonwatch.org/amazon/BR/madeira/index.php?page_number=99 (accessed April 30, 2009).

14. Danielle Knight, "Report Highlights Dams' Role in Global Warming," Inter Press Service, June 12, 2002, posted at Common Dreams website, http://www .commondreams.org/headlines02/0612–07.htm (accessed February 9, 2009).

15. "Brazil: IIRSA Madeira River Complex," Amazon Watch web site, http://www .amazonwatch.org/amazon/BR/madeira/index.php?page_number=99 (accessed April 30, 2009).

16. Knight, "Report Highlights Dams' Role in Global Warming," Inter Press Service, June 12, 2002, posted at Common Dreams website, http://www.common-dreams.org/headlines02/0612–07.htm (accessed February 9, 2009).

17. Duncan Graham-Rowe, "Hydroelectric Power's Dirty Secret Revealed," *New Scientist,* February 24, 2005, http://www.newscientist.com/article/dn7046 (accessed February 9, 2009).

18. Patrick Cunningham, "Amazon Indians Lead Battle against Power Giant's Plan to Flood Rainforest," *The Independent,* May 23, 2008, http://www.independent .co.uk/environment/climate-change/amazon-indians-lead-battle-against-power-giants-plan-to-flood-rainforest–832865.html?r=RSS (accessed April 26, 2009).

19. "Newsroom News Clip: Coordinator of Movement against Xingu Dams is Murdered," Amazon Watch, August 25, 2001, http://www.amazonwatch.org/ newsroom/view_news.php?id=93 (accessed April 26, 2009).

20. "Brazil: Belo Monte Dam," Amazon Watch, http://www.amazonwatch.org/ amazon/BR/bmd/index.php?page_number=99 (accessed April 26, 2009).

21. "Weeks Before His Death, Dema Wrote an Appeal to the World About Belo Monte Entitled 'SOS Xingu' Asking 'What Will Be Left of the Xingu River for the People of Xingu?'" Amnesty International, www.amnesty.org/en/library/asset/ AMR19/031/2001/en/e98e49ef-fb04–11dd–9fca–0d1f97c98a21/amr190312001 en.pdf (accessed April 26, 2009); Vanessa Larson, "Environmental Activists Murdered in Brazil," *World Watch,* March 1, 2002, http://www.accessmylibrary.com/ coms2/summary_0286-25110190_ITM (accessed April 26, 2009).

22. "Brazil: Belo Monte Dam," Amazon Watch web site, http://www.amazon watch.org/amazon/BR/bmd/index.php?page_number=99 (accessed April 27, 2009).

23. "Newsroom News Clip: "Coordinator of Movement against Xingu Dams is Murdered," Amazon Watch, August 25, 2001, http://www.amazonwatch.org/ newsroom/view_news.php?id=93 (accessed April 27, 2009).

24. Alan Clendenning, "Brazil All Set to Flood the Rainforest," *Scotsman,* May 23, 2008, http://thescotsman.scotsman.com/world/Brazil-all-set-to-flood.4113906 .jp (accessed February 9, 2009).

25. "Newsroom News Clip: 'Coordinator of Movement against Xingu Dams is Murdered,'" Amazon Watch, August 25, 2001, http://www.amazonwatch.org/newsroom/view_news.php?id=93 (accessed April 27, 2009).

26. Andrew Hay, "Dangerous Times on Brazil's Amazon Frontier," Reuters, December 22, 2005, http://news.mongabay.com/2005/1222-reuters.html (accessed April 27, 2009).

27. Vanessa Larson, "Environmental Activists Murdered in Brazil," World Watch, March 1, 2002, http://www.accessmylibrary.com/coms2/summary_0286–2511 0190_ITM (accessed April 27, 2009).

28. Andrew Hay, "Dangerous Times on Brazil's Amazon Frontier," Reuters, December 22, 2005, http://news.mongabay.com/2005/1222-reuters.html (accessed April 27, 2009).

29. "Weeks Before His Death, Dema Wrote an Appeal to the World About Belo Monte Entitled 'SOS Xingu' Asking 'What Will Be Left of the Xingu River for the People of Xingu?'" Amnesty International, www.amnesty.org/en/library/asset/ AMR19/031/2001/en/e98e49ef-fb04–11dd–9fca–0d1f97c98a21/amr190312001 en.pdf (accessed April 27, 2009); "Newsroom News Clip: 'Coordinator of Movement against Xingu Dams Is Murdered,'" Amazon Watch, August 25, 2001, http://www.amazonwatch.org/newsroom/view_news.php?id=93 (accessed April 27, 2009).

30. Ibid.

31. Vanessa Larson, "Environmental Activists Murdered in Brazil," World Watch, March 1, 2002, http://www.accessmylibrary.com/coms2/summary_0286–25110190_ITM (accessed April 27, 2009); "Weeks Before His Death, Dema Wrote an Appeal to the World," Amnesty International, www.amnesty.org/en/ library/asset/AMR19/031/2001/en/e98e49ef-fb04–11dd–9fca–0d1f97c98a21/ amr190312001en.pdf (accessed April 27, 2009).

32. Ibid.

33. Marcelo Ballvé, "Brazilian President's Push for Dams in Amazon Basin Stirs Controversy," World Politics Review, August 20, 2007, http://www.world politicsreview.com/Article.aspx?id=1046 (accessed February 14, 2009).

34. Ballvé, "Mega-Dams in the Amazon: Brazil's Three Gorges?" NACLA, July 26, 2007, http://nacla.org/node/1501 (accessed February 14, 2009).

35. "$11B Amazon Rainforest Dam Gets Initial Approval," mongabay.com, July 10, 2007, http://news.mongabay.com/2007/0710-madeira.html.

36. Glenn Switkes, "Re-mapping Latin America's Future (#3), Brazilian Government Moves to Dam Principal Amazon Tributary," America's Program Center for International Policy, June 12, 2007, http://americas.irc-online.org/am/4299 (accessed February 14, 2009); Marcello Ballvé, "Brazilian President's Push for Dams in Amazon Basin Stirs Controversy," World Politics Review, August 20, 2007, http://www.worldpoliticsreview.com/Article.aspx?id=1046 (accessed February 14, 2009); Marcello Ballvé, "Mega-Dams in the Amazon: Brazil's Three Gorges?" NACLA, July 26, 2007, http://nacla.org/node/1501 (accessed February 14, 2009).

37. Ballvé, "Mega-Dams in the Amazon: Brazil's Three Gorges?" NACLA, July 26, 2007, http://nacla.org/node/1501 (accessed February 14, 2009).

38. "Dead Catfish May Be the Least of Lula's Worries," International Rivers Network web site, December 23, 2008, http://www.internationalrivers.org/en/blog/ aviva-imhof/dead-catfish-may-be-least-lulas-worries (accessed February 14, 2009); "$11B Amazon Rainforest Dam Gets Initial Approval," mongabay.com,

July 10, 2007, http://news.mongabay.com/2007/0710-madeira.html (accessed April 30, 2009); Ballvé, "Mega-Dams in the Amazon: Brazil's Three Gorges?" NACLA, July 26, 2007, http://nacla.org/node/1501 (accessed February 14, 2009).

39. Greenpeace, "Slaughtering the Amazon," Part One, updated July 2009 version, http://www.greenpeace.org/international/assets/binaries/slaughtering-the-amazon-part1 (accessed July 24, 2009).

40. "The Amazon River's Largest Tributary is Under Threat," Friends of the Earth, undated report, http://www.foei.org/en/get-involved/take-action/pdfs/RIO _MADEIRA.pdf (accessed August 9, 2009).

41. Simón Farabundo Ríos, "Integration and the Environment on the Rio Madeira," NACLA, March 9, 2009, https://nacla.org/node/5595 (accessed August 9, 2009); "Unaccountable and Unsustainable, IDB Champions Corporate Interests at Expense of Citizens," Bank Information Center, April 3, 2008, http://www.bicusa .org/en/Article.3715.aspx (accessed August 9, 2009).

42. Tyler Bridges, "Will Dams on Amazon Tributary Wreak Global Havoc?," McClatchy, April 5th, 2009, posted at physorg website, http://www.physorg.com/ news158154937.html (accessed April 30, 2009).

43. Andrea Welsh, "Brazil Greens See Tensions if Lula Wins 2nd Term," Reuters News, September 28, 2006.

44. Switkes, "Re-mapping Latin America's Future (#3), Brazilian Government Moves to Dam Principal Amazon Tributary," America's Program Center for International Policy, June 12, 2007, http://americas.irc-online.org/am/4299 (accessed February 14, 2009).

45. Julie McCarthy, "Brazilian Tribes Say Dam Threatens Way of Life," National Public Radio, http://www.npr.org/templates/story/story.php?storyId=91007395 (accessed April 27, 2009).

46. Larry Rohter, "Brazil, Fearful of Blackouts, Orders 20 percent Cut in Electricity," *New York Times,* May 19, 2001, http://www.nytimes.com/2001/05/19/world/ brazil-fearful-of-blackouts-orders–20-cut-in-electricity.html?n=Top percent2F Reference percent2FTimes percent20Topics percent2FSubjects percent2FF percent2FFinances (accessed April 30, 2009).

47. Bridges, "Will Dams on Amazon Tributary Wreak Global Havoc?," McClatchy, April 5, 2009, posted at physorg website, http://www.physorg.com/news 158154937.html (accessed April 30, 2009).

48. Michael Grunwald, "The Clean Energy Scam," *Time* (Canadian Edition) 30, 171, no. 14, April 7, 2008.

49. "The Environment: Biofools," *Economist,* April 8, 2009, http://www.economist.com/science/displaystory.cfm?story_id=13437705 (accessed May 20, 2009); Deroy Murdock, "Global Food Riots: Made in Washington, D.C.," *National Review,* April 8, 2009, http://article.nationalreview.com/print/?q=OTBiOTY2Z TAyMWQwYTJkMDIwMmFiZGY4YzAxM2VkNjc= (accessed May 20, 2009).

50. "The Environment: Biofools," *Economist,* April 8, 2009, http://www.economist .com/science/displaystory.cfm?story_id=13437705 (accessed May 20, 2009).

51. Robert Bryce, "Ethanol, Food Prices and the Environment," *Washington Spectator,* August 1, 2008, http://www.washingtonspectator.org/articles/20080801 ehtanol.cfm (accessed May 20, 2009).

52. "Biofuels Could Increase Global Warming with Laughing Gas," phys.org web site, September 21, 2007, http://www.physorg.com/news109581631.html (accessed May 20, 2009).

53. Hal Bernton, "Ethanol Demand Turns Corn into a Growing Cash Crop," *Seattle Times,* November 14, 2006, http://seattletimes.nwsource.com/html/business technology/2003305110_biofuel15.html (accessed May 20, 2009).

54. Jason Beaubien, "Ethanol Demand, Prices Boost Farm Communities," National Public Radio, March 4, 2008, http://www.npr.org/templates/story/story.php ?storyId=87782087 (accessed May 20, 2009).

55. Ibid.

56. John Carey, "Controversies Continue to Swirl over Corn Ethanol," *Business Week,* April 16, 2009, http://www.businessweek.com/magazine/content/09_17/b41280 38023092.htm?campaign_id=rss_topStories/feed; Barrie McKenna, "Ethanol Helps a Small Town Bloom," *Toronto Globe and Mail,* July 25, 2007, http://www. scrippsnews.com/node/25654 (accessed May 20, 2009).

57. Beaubien, "Ethanol Demand, Prices Boost Farm Communities," National Public Radio, March 4, 2008, http://www.npr.org/templates/story/story.php?story Id=87782087 (accessed May 20, 2009).

58. "In Iowa, Questions Arise on Impact of Ethanol Production," PBS, January 28, 2009, http://www.pbs.org/newshour/bb/environment/jan-june09/mixedyield _01-28.html (accessed May 20, 2009).

59. "Study Critiques Corn-for-ethanol's Carbon Footprint," phys.org website, March 2, 2009, http://www.physorg.com/news155229840.html (accessed May 20, 2009).

60. "In Iowa, Questions Arise on Impact of Ethanol Production," PBS, January 28, 2009, http://www.pbs.org/newshour/bb/environment/jan-june09/mixedyield _01-28.html (accessed May 20, 2009).

61. "The Environment: Biofools," *Economist,* April 8, 2009, http://www.economist .com/science/displaystory.cfm?story_id=13437705 (accessed May 20, 2009).

62. "Biofuels Could Increase Global Warming with Laughing Gas," phys.org website, September 21, 2007, http://www.physorg.com/news109581631.html (accessed May 20, 2009).

63. Luciana Pereira Franco, "In Brazil, Biofuels Dream Is Already Reality," *Washington Post,* October 29, 2008, http://www.washingtonpost.com/wp-dyn/ content/article/2008/10/28/AR2008102801368.html (accessed May 20, 2009).

64. Nicole Gaouette and Richard Simon, "High Food Prices May Put Farmers on a Subsidy Diet," *Los Angeles Times,* May 2, 2008, http://articles.latimes.com/2008/ may/02/nation/na-foodprices2 (accessed May 20, 2009).

65. "Congressional Scorecard 2006," Republicans for Environmental Protection, http://www.rep.org/2006_scorecard.pdf (accessed May 20, 2009); "Congressional Scorecard 2007," Republicans for Environmental Protection, http://www.rep.org/ 2007_scorecard.pdf (accessed May 20, 2009).

66. Matt Kibbe, "Chuck Grassley, King of Pork," *Washington Spectator,* June 10, 2008, http://spectator.org/archives/2008/06/10/chuck-grassley-king-of-pork (accessed May 20, 2009).

67. Sheryl Gay Stolberg, "Ease a Little Guilt, Provide Some Jobs: It's Pork on the Hill," *New York Times,* December 20, 2003, http://www.nytimes.com/2003/12/ 20/us/ease-a-little-guilt-provide-some-jobs-it-s-pork-on-the-hill.html?page wanted=all (accessed May 20, 2009).

68. "Coralville City Leaders Monitoring Rainforest Project," *Sioux City Journal,* April 12, 2005, http://www.siouxcityjournal.com/articles/2005/04/12/news/latest _news/98fbbf2b2780cd9086256fe1005599ad.txt (accessed May 20, 2009).

69. Alan Scher Zagier, "If They Build It, Who Will Come? Iowa Builders Tout $180m Indoor Rain Forest, but Critics Unconvinced," *Boston Globe,* January 8, 2005 http://www.boston.com/news/nation/articles/2005/01/08/if_they_build _it_who_will_come/ (accessed May 20, 2009).

70. "Frequently Asked Questions," Earthpark web site, http://www.earthpark.net/ about/faqs.php (accessed May 20, 2009).

71. Zagier, "If They Build It, Who Will Come?," *Boston Globe,* January 8, 2005, http://www.boston.com/news/nation/articles/2005/01/08/if_they_build_it_who _will_come/ (accessed May 20, 2009).

72. "Earthpark Planned for Iowa," October 10, 2006, *USA Today,* http://www.usa today.com/travel/destinations/2006–10–10-iowa-earthpark_x.htm (accessed May 20, 2009).

73. Zagier, "If They Build It, Who Will Come?," *Boston Globe,* January 8, 2005, http://www.boston.com/news/nation/articles/2005/01/08/if_they_build_it_who _will_come/ (accessed May 20, 2009).

74. Michael Judge, "The Incredible Shrinking Rain Forest," *Wall Street Journal,* March 9, 2006, http://online.wsj.com/article/SB114187377486193364-search .html (accessed May 20, 2009).

75. "In Iowa, Questions Arise on Impact of Ethanol Production," PBS, January 28, 2009, http://www.pbs.org/newshour/bb/environment/jan-june09/mixedyield_ 01–28.html (accessed May 20, 2009).

76. "Nature Conservancy Study Raises Major Questions on Biofuels," *Biotech Week* 3776, February 27, 2008.

77. Sabrina Valle, "Sugar Cane Eats into Land / World's Appetite for Ethanol Affects a Region of Brazil," *Washington Post,* August 3, 2007.

78. Grunwald, "The Clean Energy Scam," *Time* (Canadian Edition) 30, 171, no. 14, April 7, 2008.

79. Kenneth Rapoza, "Brazil Soy Industry Prepares for Biodiesel War with Argentina," Market Watch, March 25, 2007, http://www.marketwatch.com/news/ story/brazil-soy-industryprepares-biodiesel-war/story.aspx?guid={25F4E372– 55D8–41D4–9E71–92A4031C0088} (accessed May 20, 2009).

80. "Brazil Increases Consumption, Export of Sugar-Based Ethanol," Voice of America Press Releases and Documents, August 14, 2008.

81. "Brazil's Biofuel Plane Fleet Grows," Agence France Presse, http://afp.google .com/article/ALeqM5jEI9bv22bR0ebv6090PbSFFsx6Iw (accessed May 20, 2009).

82. Lenilson Ferreira, "Brazil Priming Ethanol Initiative to Supply Fuel-thirsty Japan," Kyodo, August 22, 2007, http://search.japantimes.co.jp/cgi-bin/nb2007 0822a3.html (accessed May 20, 2009).

83. Joshua Schneyer, "The Money Flying Down to Brazil; Big-name Investors Sniff Hefty Ethanol Profits in the Country's Vast Cane Fields," *Business Week* 70, Volume 4039, June 18, 2007.

84. Franco, "In Brazil, Biofuels Dream Is Already Reality," *Washington Post,* October 29, 2008, http://www.washingtonpost.com/wp-dyn/content/article/2008/10/28/ AR2008102801368.html (accessed May 20, 2009).

85. Chris McGowan, "Biofuel Could Eat Brazil's Savannas & Deforest the Amazon," Huffington Post, September 14, 2007, http://www.huffingtonpost.com/chris-mcgowan/biofuel-could-eat-brazil_b_64466.html (accessed May 20, 2009).

86. Jan Rocha, "Society: The Sweet Hereafter: Biofuels Are Now Seen as Polluting and as a Threat to Forests and Food Production," *Guardian,* May 28, 2008.

87. Virginia Morell, "The Rain Forest in Rio's Backyard," *National Geographic,* March 2004, http://environment.nationalgeographic.com/environment/habitats/rio-rain-forest.html (accessed May 20, 2009).

88. McGowan, "Biofuel Could Eat Brazil's Savannas & Deforest the Amazon," Huffington Post, September 14, 2007, http://www.huffingtonpost.com/chris-mcgowan/biofuel-could-eat-brazil_b_64466.html (accessed May 20, 2009).

89. Morell, "The Rain Forest in Rio's Backyard," *National Geographic,* March 2004, http://environment.nationalgeographic.com/environment/habitats/rio-rain-forest.html (accessed May 20, 2009).

90. Terry Macalister, "Biofuels: Brazil Disputes Cost of Sugar in the Tank: Ethanol Producers Cut the Carbon but Still Come Under Fire from Campaigners," *The Guardian,* June 10, 2008; Morell, "The Rain Forest in Rio's Backyard," *National Geographic,* March 2004, http://environment.nationalgeographic.com/environment/habitats/rio-rain-forest.html (accessed May 20, 2009).

91. Macalister, "Biofuels: Brazil Disputes Cost of Sugar in the Tank: Ethanol Producers Cut the Carbon but Still Come Under Fire from Campaigners," *The Guardian,* June 10, 2008.

92. "Brazil: Places We Protect, The Atlantic Forest," Nature Conservancy, http://www.nature.org/wherewework/southamerica/brazil/work/art5080.html (accessed May 20, 2009).

93. McGowan, "Biofuel Could Eat Brazil's Savannas & Deforest the Amazon," Huffington Post, September 14, 2007, http://www.huffingtonpost.com/chris-mcgowan/biofuel-could-eat-brazil_b_64466.html (accessed May 20, 2009).

94. "Brazil Fines 24 Ethanol Producers for Illegal Forest Clearing," mongabay.com, July 1, 2008, http://news.mongabay.com/2008/0701-brazil.html (accessed May 20, 2009).

95. Jim Lane, "Brazil Fines 24 Ethanol Producers $75 Million for Atlantic Rainforest Destruction," *Biofuels Digest,* July 2, 2008, http://www.biofuelsdigest.com/blog2/2008/07/02/brazil-fines-24-ethanol-producers-75-million-for-atlantic-rainforest-destruction/ (accessed May 20, 2009).

96. "Brazil: Places We Protect, the Atlantic Forest," Nature Conservancy, http://www.nature.org/wherewework/southamerica/brazil/work/art5080.html (accessed May 20, 2009).

97. Morell, "The Rain Forest in Rio's Backyard," *National Geographic,* March 2004, http://environment.nationalgeographic.com/environment/habitats/rio-rain-forest.html (accessed May 20, 2009).

98. James Brooke, "Brazilian Rain Forest Yields Most Diversity for Species of Trees," *New York Times,* March 30, 1993, http://www.nytimes.com/1993/03/30/science/brazilian-rain-forest-yields-most-diversity-for-species-of-trees.html (accessed May 20, 2009).

99. "Brazil: Places We Protect, The Atlantic Forest," Nature Conservancy, http://www.nature.org/wherewework/southamerica/brazil/work/art5080.html (accessed May 20, 2009).

100. "Brazil fines 24 ethanol producers for illegal forest clearing," mongabay.com, July 1, 2008, http://news.mongabay.com/2008/0701-brazil.html (accessed May 20, 2009).

101. "Golden Lion Tamarin Conservation Program," Smithsonian National Zoological Park, http://nationalzoo.si.edu/ConservationAndScience/Endangered-Species/GLTProgram/Learn/BasicFacts.cfm (accessed May 20, 2009).

102. "Maned Three-toed Sloth Bradypus torquatus," World Land Trust, http://www.worldlandtrust.org/animals/maned-sloth.htm (accessed May 20, 2009); Adriano G. Chiarello, "Diet of the Atlantic Forest Maned Sloth Bradypus torquatus (Xenarthra: Bradypodidae)," abstract, *Journal of Zoology* 246 (1998), 11–19, Cambridge University Press, http://journals.cambridge.org/action/displayAbstract;jsessionid=11F2855608B9ABE034E894D2C42E56B2.tomcat1?fromPage=online&aid=41169 (accessed May 20, 2009).

103. Ibid.

104. "Maned Sloth–Bradypus torquatus," United Nations Environment Program, http://www.unep-wcmc.org/species/data/species_sheets/manedslo.htm (accessed May 20, 2009).

105. "Maned Three-toed Sloth Bradypus torquatus," World Land Trust, http://www.worldlandtrust.org/animals/maned-sloth.htm (accessed May 20, 2009).

106. "Brazil's Ethanol Push Could Eat Away at Amazon," MSNBC, March 7, 2007 http://www.msnbc.msn.com/id/17500316/page/2/ (accessed May 20, 2009).

107. Tom Phillips, "Brazil's Ethanol Slaves," *Guardian,* March 9, 2007, http://www.commondreams.org/headlines07/0309–08.htm (accessed May 20, 2009).

108. Barbara J. Fraser, "Agri-Bonanza's Ripples—Church Cautiously Watches Ethanol Boom," Catholic News Service, August 14, 2007, http://www.catholicnews.com/data/stories/cns/0704627.htm (accessed October 10, 2009).

109. Clemens Höges, "A 'Green Tsunami' in Brazil, the High Price of Clean, Cheap Ethanol," *Der Spiegel,* http://www.spiegel.de/international/world/0,1518,60295 1,00.html (accessed May 20, 2009).

110. Phillips, "Brazil's Ethanol Slaves," *The Guardian,* March 9, 2007, http://www.commondreams.org/headlines07/0309–08.htm (accessed May 20, 2009).

111. Höges, "A 'Green Tsunami' in Brazil," *Der Spiegel,* http://www.spiegel.de/international/world/0,1518,602951,00.html (accessed May 20, 2009).

112. V. Reis, S. Lee, and C. Kennedy, "Biological Nitrogen Fixation in Sugarcane," in Claudine Elmerich, *Associative and Endophytic Nitrogen-fixing Bacteria and Cyanobacterial Associations* (New York: Springer, 2006), 214, Google Books, http://books.google.com/books?id=hiEkEBI1-Y8C&pg=PA214&lpg=PA214&dq=nitrogen+fertilizer+ethanol+brazil&source=bl&ots=fPKAwMvwm8&sig=ObjRDUNb3p_7lsSyh2h0vjd8HgA&hl=en&ei=0msGSsrnOM6Ltgf1_p2TBw&sa=X&oi=book_result&ct=result&resnum=9#PPA214,M1 (accessed May 20, 2009).

113. "Cerrado Deforestation Contributes to Global Warming, Specialists Say," Agencia Senado, September 14, 2007, http://www.senado.gov.br/agencia/internacional/en/not_367.aspx (accessed May 20, 2009).

114. Jack Chang, "Cane Catches Fire; Brazil Adapts an Old Crop to the Raging Biofuels Market," *Financial Post,* May 21, 2008.

115. Höges, "A 'Green Tsunami' in Brazil," *Der Spiegel,* http://www.spiegel.de/international/world/0,1518,602951,00.html (accessed May 20, 2009).

116. Mario Osava, "Energy: Brazil Aims to Dominate World Ethanol Market," Tierramérica, March 31, 2007, http://ipsnews.net/news.asp?idnews=37172 (accessed May 20, 2009); Joel Bourne, "Green Dreams," *National Geographic,* October 2007, http://ngm.nationalgeographic.com/ngm/2007–10/biofuels/biofuels-p4.html (accessed May 20, 2009).

117. Ferreira, "Brazil Priming Ethanol Initiative to Supply Fuel-thirsty Japan," Kyodo, August 22, 2007, http://search.japantimes.co.jp/cgi-bin/nb20070822a3.html (accessed May 20, 2009).

118. McGowan, "Biofuel Could Eat Brazil's Savannas & Deforest the Amazon," Huffington Post, September 14, 2007, http://www.huffingtonpost.com/chris-mcgowan/biofuel-could-eat-brazil_b_64466.html (accessed May 20, 2009).

119. Macalister, "Biofuels: Brazil Disputes Cost of Sugar in the Tank: Ethanol Producers Cut the Carbon but Still Come Under Fire from Campaigners," *Guardian*, June 10, 2008.

120. Alan Clendenning, "Brazil: Ethanol Farming Won't Impact Amazon Rain Forest," Associated Press, Jul 10, 2007, http://findarticles.com/p/articles/mi_qn4176/is_20070710/ai_n19354074/ (accessed May 20, 2009); Reese Ewing, "Brazil Fends Off Critics of Its Biofuel Proponents of Cane-based Ethanol Dismiss Attacks as Myths," Reuters, May 14, 2008.

121. "Sugarcane Pushing Brazil Cattle North Near Amazon—Estado," Dow Jones International News, June 30, 2008.

122. Macalister, "Biofuels: Brazil Disputes Cost of Sugar in the Tank: Ethanol Producers Cut the Carbon but Still Come Under Fire from Campaigners," *Guardian*, June 10, 2008.

123. Clendenning, "Brazil: Ethanol Farming Won't Impact Amazon Rain Forest," *Oakland Tribune*, July 10, 2007, http://findarticles.com/p/articles/mi_qn4176/is_20070710/ai_n19354074/ (accessed May 20, 2009).

124. Fraser, "Agri-Bonanza's Ripples," Catholic News Service, August 14, 2007, http://www.catholicnews.com/data/stories/cns/0704627.htm (accessed October 10, 2009).

125. "CNAA Closes on IDB Debt," Project Finance, February 23, 2009, http://www.projectfinancemagazine.com/default.asp?page=7&PubID=4&ISS=25297&SID=717588 (accessed October 10, 2009).

126. Antonio Regalado, "Development Bank to Make $269 Million Loan to Brazil Ethanol Maker," *Wall Street Journal*, July 24, 2008, http://online.wsj.com/article/SB121686052327479467.html?mod=googlewsj (accessed August 9, 2009); Robert Scheer, "Making Money, the Bush Family Way," Salon.com, February 20, 2002, http://dir.salon.com/story/news/col/scheer/2002/02/20/carlyle/index.html (accessed August 9, 2009).

127. Jamie Doward, "'Ex-presidents Club' Gets Fat on Conflict," *Observer*, Sunday, March 23, 2003, http://www.guardian.co.uk/business/2003/mar/23/iraq.the observer (accessed August 9, 2009); Scheer, "Making Money, the Bush Family Way," Salon.com, February 20, 2002, http://dir.salon.com/story/news/col/scheer/2002/02/20/carlyle/index.html (accessed August 9, 2009); "About 100 arrested at protest outside Manhattan investment house," CNN, April 7, 2003, http://edition.cnn.com/2003/US/Northeast/04/07/ny.protest/ (accessed August 9, 2009).

128. Stephen Foley, "Abu Dhabi buys 7.5 per Cent Stake in Carlyle," *Independent*, September 21, 2007, http://www.independent.co.uk/news/business/news/abu-dhabi-buys–75-per-cent-stake-in-carlyle–403063.html (accessed August 9, 2009).

129. McGowan, "Biofuel Could Eat Brazil's Savannas & Deforest the Amazon," Huffington Post, September 14, 2007, http://www.huffingtonpost.com/chris-mcgowan/biofuel-could-eat-brazil_b_64466.html (accessed May 20, 2009).

130. Hal Bernton, "Ethanol Demand Turns Corn into a Growing Cash Crop," *Seattle Times,* November 14, 2006, http://seattletimes.nwsource.com/html/business technology/2003305110_biofuel15.html (accessed May 20, 2009).

131. David Adams, "Jeb Bush Encouraged Brother to Pursue Ethanol," *St. Petersburg Times,* March 5, 2007, http://www.sptimes.com/2007/03/05/Worldandnation/ Jeb_Bush_encouraged_b.shtml (accessed May 20, 2009).

132. Mario Osava, "Energy: Brazil Aims to Dominate World Ethanol Market," Tierr américa, March 31, 2007, http://ipsnews.net/news.asp?idnews=37172 (accessed May 20, 2009).

133. "Brazil: Government Grants Temporary Approval fro Genetically Modified Soy," NotiSur–South American Political and Economic Affairs, Latin American Data Base/Latin American Institute, October 3, 2003, http://www.accessmylibrary .com/coms2/summary_0286–24567050_ITM (accessed April 12, 2009); "Prize recognizes largest contributor to Amazon rainforest destruction," mongabay.com, May 27, 2005, http://news.mongabay.com/2005/0527-golden_chainsaw.html (accessed April 12, 2009); Adams, "Jeb Bush Encouraged Brother to Pursue Ethanol," *St. Petersburg Times,* March 5, 2007, http://www.sptimes.com/2007/ 03/05/Worldandnation/Jeb_Bush_encouraged_b.shtml (accessed May 20, 2009).

134. Michael Smith and Carlos Caminada, "Ethanol's Deadly Brew," Bloomberg Markets, November 2007, http://www.bloomberg.com/news/marketsmag/mm _1107_story3.html (accessed May 20, 2009).

135. "Ethanol Commission Launched in Miami," The Fueling Station, December 18, 2006, http://blogs.tampabay.com/energy/2006/12/ethanol_commiss.html (accessed May 20, 2009).

136. "Biofuels, Agrofuels, the International Financial Institutions and Private Investment: A General Panorama," Bank Information Center, October 3, 2008, http:// www.bicusa.org/en/Article.10385.aspx (accessed May 20, 2009).

137. Smith and Carlos Caminada, "Ethanol's Deadly Brew," Bloomberg Markets, November 2007, http://www.bloomberg.com/news/marketsmag/mm_1107_story3 .html (accessed May 20, 2009); Adams, "Jeb Bush Encouraged Brother to Pursue Ethanol," *St. Petersburg Times,* March 5, 2007, http://www.sptimes.com/ 2007/03/05/Worldandnation/Jeb_Bush_encouraged_b.shtml (accessed May 20, 2009).

138. Glenn Hurowitz, "George Soros vs. the Planet," Huffington Post, August 7, 2007, http://www.huffingtonpost.com/glenn-hurowitz/george-soros-vs-the-plan _b_59502.html (accessed May 20, 2009).

139. Jan Rocha, "Society: The Sweet Hereafter: Biofuels Are Now Seen as Polluting and as a Threat to Forests and Food Production," *Guardian,* May 28, 2008.

140. McGowan, "Biofuel Could Eat Brazil's Savannas & Deforest the Amazon," Huffington Post, September 14, 2007, http://www.huffingtonpost.com/ chris-mcgowan/biofuel-could-eat-brazil_b_64466.html (accessed May 20, 2009).

141. Daniel Howden, "Brazil's Experience Testifies to the Downside of this Energy Revolution," *Independent,* April 15, 2008.

142. "False Claims of the Agrofuels Lobby," Friends of the Earth Europe, http://www .foeeurope.org/corporates/worstlobby/false_claims_agrofuels.pdf (accessed May 20, 2009).

143. Ferreira, "Brazil Priming Ethanol Initiative to Supply Fuel-thirsty Japan," Kyodo, August 22, 2007, http://search.japantimes.co.jp/cgi-bin/nb20070822a3.html (accessed May 20, 2009).

144. McGowan, "Biofuel Could Eat Brazil's Savannas & Deforest the Amazon," Huffington Post, September 14, 2007, http://www.huffingtonpost.com/chris-mcgowan/biofuel-could-eat-brazil_b_64466.html (accessed May 20, 2009).

145. "Ethanol Escapade: Brazil Sugar Producers, Criticized, Seek Sweeter PR," Wall Street Journal, June 13, 2008, http://blogs.wsj.com/environmentalcapital/2008/06/13/ethanol-escapade-brazil-sugar-producers-criticized-seek-sweeter-pr/ (accessed May 20, 2009); "High-Ranking US and Brazilian Officials Defend Biofuel Programs," The Oil Daily, May 22, 2008; "Brazil: Biofuel Backlash," Energy Compass, May 30, 2008.

146. Grunwald, "The Clean Energy Scam," Time (Canadian Edition) 30, Volume 171, no. 14, April 7, 2008.

147. Rocha, "Society: The Sweet Hereafter: Biofuels Are Now Seen as Polluting and as a Threat to Forests and Food Production," The Guardian, May 28, 2008.

148. Terry Macalister, "Biofuels: Brazil Disputes Cost of Sugar in the Tank: Ethanol Producers Cut the Carbon but Still Come Under Fire from Campaigners," Guardian, June 10, 2008.

149. "Brazil's Ethanol Push Could Eat Away at Amazon," MSNBC, March 7, 2007, http://www.msnbc.msn.com/id/17500316/page/2/ (accessed May 20, 2009).

150. "Lula says Brazil Ready to Share Sugarcane Biofuel Technology," Xinhua, April 19, 2009, http://news.xinhuanet.com/english/2009-04/19/content_11212325.htm (accessed May 20, 2009).

151. Andrew Downie, "Brazil Defends Ethanol in Food-versus-fuel Fight," Christian Science Monitor, May 5, 2008, http://www.csmonitor.com/2008/0505/p04s01-woam.html (accessed May 20, 2009).

152. McGowan, "Biofuel Could Eat Brazil's Savannas & Deforest the Amazon," Huffington Post, September 14, 2007, http://www.huffingtonpost.com/chris-mcgowan/biofuel-could-eat-brazil_b_64466.html (accessed May 20, 2009).

153. "Transcript: David Brancaccio interviews George Soros," September 12, 2003, PBS, http://www.pbs.org/now/transcript/transcript_soros.html (accessed May 20, 2009).

154. McGowan, "Biofuel Could Eat Brazil's Savannas & Deforest the Amazon," Huffington Post, September 14, 2007, http://www.huffingtonpost.com/chris-mcgowan/biofuel-could-eat-brazil_b_64466.html (accessed May 20, 2009); Rachel Oliver, "All About: Eco-philanthropy," CNN, March 2, 2008, http://www.cnn.com/2008/BUSINESS/03/02/eco.philanthropy/ (accessed May 20, 2009).

155. Carl Mortished, "Billionaire Speculators Fuel Drive for Petrol Alternative," Times (of London), June 9, 2007, http://business.timesonline.co.uk/tol/business/industry_sectors/natural_resources/article1906991.ece (accessed May 20, 2009).

156. Franco, "In Brazil, Biofuels Dream Is Already Reality," Washington Post, October 29, 2008, http://www.washingtonpost.com/wp-dyn/content/article/2008/10/28/AR2008102801368.html (accessed May 20, 2009).

157. Isabella Kenfield, "Americas Program Discussion Paper: Fueling the Debate: Agrofuels, Biodiversity, and Our Energy Future (#2), Brazil's Ethanol Plan Breeds Rural Poverty, Environmental Degradation," Americas Program, Center For International Policy, March 6, 2007, http://americas.irc-online.org/am/4049 (accessed May 20, 2009); Isabella Kenfield and Roger Burbach, "Militant Brazilian Opposition to Bush-Lula Ethanol Accords," Transnational Institute, March 21, 2007, http://www.tni.org/detail_page.phtml?act_id=16499 (accessed May 20, 2009).

158. Peter Baker and Bill Brubaker, "Bush Hails International Ethanol Production," *Washington Post,* March 9, 2007, http://www.washingtonpost.com/wp-dyn/content/article/2007/03/09/AR2007030900767_pf.html (accessed August 9, 2009).

159. Fabiana Frayssinet, "Agriculture Brazil: David, Goliath and Land Reform," Inter Press Service, June 13, 2007, http://ipsnews.net/news.asp?idnews=38161 (accessed May 20, 2009).

160. Kenfield and Burbach, "Militant Brazilian Opposition to Bush-Lula Ethanol Accords," Transnational Institute, March 21, 2007, http://www.tni.org/detail_page.phtml?act_id=16499 (accessed May 20, 2009).

161. "Greenpeace Protest Greets Lula-Bush Biofuels Meeting in São Paulo," Greenpeace, March 8, 2007, http://www.greenpeace.org/usa/press-center/releases2/greenpeace-protest-greets-lula (accessed May 20, 2009).

162. Isabella Kenfield and Roger Burbach, "Militant Brazilian Opposition to Bush-Lula Ethanol Accords," Transnational Institute, March 21, 2007, http://www.tni.org/detail_page.phtml?act_id=16499 (accessed May 20, 2009).

163. "Methane From Dams: Greenhouse Gas to Power Source," Environmental News Service, May 9, 2007, http://www.ens-newswire.com/ens/may2007/2007–05–09–04.asp (accessed February 9, 2009).

164. Sheila McNulty, "US Wind Power: Industry Sets Sights on Large Slice of the Pie," *Financial Times,* May 26, 2009, http://www.ft.com/cms/s/393c5164–48c5–11de–8870–00144feabdc0.html (accessed June 3, 2009).

165. Les Blumenthal, "Obama Seeks Funding Cuts for Wave, Tidal Energy Research," McClatchy Newspapers, http://www.kansascity.com/437/story/1225564.html (accessed June 3, 2009).

166. McGowan, "Biofuel Could Eat Brazil's Savannas & Deforest the Amazon," Huffington Post, September 14, 2007, http://www.huffingtonpost.com/chris-mcgowan/biofuel-could-eat-brazil_b_64466.html (accessed May 27, 2009).

167. "Climate Change Summit Opens with Call for Global Solidarity," *Economic Times,* December 2, 2008, http://economictimes.indiatimes.com/ET_Cetera/Climate_change_meet_for_global_solidarity/articleshow/3782978.cms (accessed August 9, 2009).

168. Martin Khor, "Obama, Climate Change and China," *Malaysia Star* online, November 10, 2008, http://thestar.com.my/news/story.asp?file=/2008/11/10/columnists/globaltrends/2503386&sec=globaltrends (accessed August 9, 2009).

169. F. James Sensenbrenner Jr., "Protecting Energy and Technology Breakthroughs," *Forbes,* June 23, 2009, http://www.forbes.com/2009/06/23/china-patent-protections-opinions-contributors-sensenbrenner.html (accessed August 9, 2009).

170. Stephen Leahy, "Environment: Climate Pact Is No Kyoto, Experts Say," Interpress Service, July 29, 2005, http://www.ipsnews.net/news.asp?idnews=29716 (accessed August 9, 2009).

171. Jeremy Lovell, "Interveiw—UN Climate Head Welcomes Marshall Plan Climate Fund," Reuters, January 16, 2008, http://www.alertnet.org/thenews/newsdesk/L16310023.htm (accessed August 9, 2009).

172. Abid Aslam, "Climate Change: Greens, Lawmakers Assail World Bank Funds," Inter Press Service, June 6, 2008, http://www.ipsnews.net/news.asp?idnews=42687 (accessed August 9, 2009).

173. Shaun Tandon, "US Draws Line with China on Climate Technology," *Sydney Morning Herald,* June 23, 2009, http://news.smh.com.au/breaking-news-

technology/us-draws-line-with-china-on-climate-technology–20090623-
cuqr.html (accessed August 9, 2009).

174. Ramesh Jaura, "Climate Change: Obama Sounds Too Much Like Bush," Inter Press Service, June 12, 2009, http://www.ipsnews.net/news.asp?idnews=47202 (accessed August 9, 2009).

175. Tandon, "US Draws Line with China on Climate Technology," *Sydney Morning Herald*, June 23, 2009, http://news.smh.com.au/breaking-news-technology/us-draws-line-with-china-on-climate-technology–20090623-cuqr.html (accessed August 9, 2009).

176. "World Emerging Economies Agree on Warming Limit, Fail on Emission Cut Pledge," Xinhua, July 10, 2009, http://news.xinhuanet.com/english/2009–07/10/content_11682838.htm (accessed August 9, 2009).

177. Khor, "Obama, Climate Change and China," *Malaysia Star* online, November 10, 2008, http://thestar.com.my/news/story.asp?file=/2008/11/10/columnists/globaltrends/2503386&sec=globaltrends (accessed August 9, 2009).

EPILOGUE

1. Sebastian Rotella, "'Drought Industry' Feeds a Hunger Crisis," *Los Angeles Times*, July 18, 1998, http://articles.latimes.com/1998/jul/18/news/mn–4791 (accessed January 16, 2009).

2. Roger Burbach, "Brazil's Landless Take the Law," Pacific News, July 6, 1998, http://www.pacificnews.org/jinn/stories/4.14/980706-brazil.html (accessed January 16, 2009).

3. Rotella, "'Drought Industry' Feeds a Hunger Crisis," *Los Angeles Times*, July 18, 1998, http://articles.latimes.com/1998/jul/18/news/mn–4791 (accessed January 16, 2009).

4. Burbach, "Brazil's Landless Take the Law," Pacific News, July 6, 1998, http://www.pacificnews.org/jinn/stories/4.14/980706-brazil.html (accessed January 16, 2009).

5. Rotella, "'Drought Industry' Feeds a Hunger Crisis," *Los Angeles Times*, July 18, 1998, http://articles.latimes.com/1998/jul/18/news/mn–4791 (accessed January 16, 2009).

6. "The President's news conference with President Cardoso in Brasilia," Weekly Compilation of Presidential Documents, October 20, 1997, http://findarticles.com/p/articles/mi_m2889/is_n42_v33/ai_20418706/pg_1?tag=artBody;col1 (accessed January 16, 2009).

7. Mario Osava, "BRAZIL: Rainwater Tanks a Weapon Against Drought," Inter Press Service, December 12, 2006, http://ipsnews.net/news.asp?idnews=35818 (accessed January 15, 2009).

8. Catarina Chagas, "Global Warming to 'Change Face of Brazilian Farming,'" SciDev.Net website (nonprofit organization providing scientific information on the developing world), August 27, 2008, http://www.scidev.net/en/climate-change-and-energy/climate-change-in-brazil/news/global-warming-to-change-face-of-brazilian-farming.html (accessed January 30, 2009).

9. Rodney Cooke and James Cock, "Cassava crops up again," *New Scientist*, June 17, 1989, http://www.newscientist.com/article/mg12216694.700—cassava-crops-up-again-cassava-is-a-buttress-against-starvation-in-many-countries—

its-starchy-roots-thrive-in-poor-soils-with-little-rainfall—but-harvested-cassava-rots-quickly-and-can-poison-people-.html (accessed February 10, 2009).

10. Interview with Miguel Pinedo Vazquez, New York, NY, September 4, 2008.
11. Interview with Javier Noriega, Iquitos October 20, 2008.
12. Interview with Noriega, Iquitos, Peru, October 20, 2008.
13. Ibid.
14. Ibid.
15. Ibid.
16. Interview with Juan Berchota, Iquitos October 17, 2008.
17. Interview with Fernando Rodríguez, Iquitos, Peru, October 17, 2008.
18. Rhys Blakely and Shivani Khanna, "Climate Change Blamed as Mango Harvest Goes Sour in India," *Times* (of London), June 9, 2008, http://www.timesonline .co.uk/tol/news/environment/article4092866.ece (accessed January 26, 2009).
19. Neva Beach, *The Ghirardelli Chocolate Cookbook: Recipes and Love from One of Americas Oldest Chocolate Companies* (Berkeley: Ten Speed Press, 1995), 4, Google Books, http://books.google.com/books?id=x2XtXQhgrlcC&pg=PA4&lpg=PA4 &dq=james+lick+ghirardelli&source=bl&ots=BWOhKMAPk8&sig= KLYcMMf4dW9thvLaCOZJgBsZyuU&hl=en&ei=r5GMSZn8D5bYygWP97 W_Bg&sa=X&oi=book_result&resnum=3&ct=result#PPA4,M1 (accessed February 6, 2009).
20. "Climate change endangers global chocolate supply," Ghana Business News, December 19, 2008, http://ghanabusinessnews.com/2008/12/19/climate-change-endangers-global-chocolate-supply/ (accessed January 28, 2009).
21. "Cambio climático afecta severamente productividad de la Amazonía," Coordinadora Nacional de Radio, May 14, 2008, http://www.cnr.org.pe/noticia.php ?id=22084 (accessed January 8, 2009); SIXTO FLORES SANCHO, "ANÁLISIS DE LOS "FRIAJES" EN LA AMAZONÍA PERUANA DURANTE ELMES DE JULIO DEL 2000," *Revista de trabajos de investigación. CNDG— Sismología, Instituto Geofísico del Perú.* vol. 2 (2001): 21 –30, posted at Geophysical Institute of Peru website, http://khatati.igp.gob.pe/cns/servicios/revista _2000/pdf/sixto_clima_final.pdf (accessed August 9, 2009).
22. Dean Cycon, "Will Coffee Be a Casualty of Climate Change?" Alternet, January 25, 2008, http://www.alternet.org/environment/74602/ (accessed January 28, 2009); Dana Ford, "Global Warming Pushes Peru," Reuters, August 20, 2008, http://www .reuters.com/article/rbssConsumerGoodsAndRetailNews/idUSN13373023200808 20?pageNumber=2&virtualBrandChannel=0 (accessed January 26, 2009).
23. Cycon, "Will Coffee Be a Casualty?" Alternet, January 25, 2008, http://www .alternet.org/environment/74602/ (accessed January 28, 2009).
24. Néfer Muñoz, "New Outbreaks of Hunger in Latin America," Tierramérica, undated, 2002, http://www.tierramerica.net/english/2002/0609/iarticulo.shtml (accessed February 11, 2009)
25. "Discussing, but Not Seeing, Poverty at Lima Summit," EU Business, May 16, 2008, http://www.eubusiness.com/news-eu/1210960051.99/ (accessed February 11, 2009).
26. "We Are Facing a Humanitarian Drama of Incalculable Consequences," speech by Esteban Lazo Hernández, vice president of the Council of State of the Republic of Cuba, at the Presidential Summit on Food Sovereignty and Security for Life, published in *Granma,* May 7, 2008, http://www.granma.cu/ingles/2008/ mayo/juev8/20ldrama.html (accessed February 12, 2009).

27. "Global Warming Hits Islands," Environment News Service, June 4, 2001, posted at Wired website, http://www.wired.com/techbiz/media/news/2001/06/44283 (accessed August 9, 2009).
28. José Adán Silva, "Latin America: Food Summit Declares Regional Emergency," Inter Press Service, May 8, 2008, http://ipsnews.net/news.asp?idnews=42294 (accessed January 30, 2009).

CONCLUSION

1. Silberner, "Climate Connections," NPR Morning Edition, http://www.npr.org/templates/story/story.php?storyId=16354380 (accessed June 3, 2009) This book has criticized the many environmental shortcomings of the Lula administration. However during climate negotiations in Copenhagen, Brazil pledged to cut its greenhouse gas emissions between 36.1% and 38.9%, largely by controlling deforestation in the Amazon. Since this book went to press, Lula signed a law requiring Brazil to reduce its emissions by said targets despite the failure of the Copenhagen summit to establish binding limits on emissions. Though environmentalists are skeptical that the government will comply with the legislation, the new law sets a positive example for other nations. "Brazil to Keep Emissions Reductions Pledge Despite Failed Climate Summit," Mongobay.com, November 15, 2009; "Lula OKs law for 40% Emission Cut by 2020," Agence France Presse, December 31, 2009; "Greenpeace Acusa a Lula de tener un doble discuso en politica mediaambiental," EFEC (Spanish news agency), December 29, 2009.
2. Dennis del Castillo Torres, Erasmo Otárola Acevedo, and Luis Freitas Alvarado, "The Amazing Palm Tree of the Amazon," Instituto de Investigaciones de la Amazonía Peruana website, http://www.iiap.org.pe/Publicaciones/CD/documentos/L028.pdf (accessed June 3, 2009).
3. "Correa calificó de 'hipocresía internacional' la falta de financiación para ITT," EFE news agency, March 10, 2009, posted at *El Comercio* website, http://www.elcomercio.com/solo_texto_search.asp?id_noticia=169282&anio=2009&mes=3&dia=10 (accessed May 27, 2009).
4. "Biochar and Reforestation May Offer Better Global Cooling Potential than Ccean Fertilization," mongabay.com, January 28, 2009, http://news.mongabay.com/2009/0128-geoengineering.html (accessed May 27, 2009).
5. Cycon, "Will Coffee Be a Casualty?" Alternet, January 25, 2008, http://www.alternet.org/environment/74602/ (accessed January 28, 2009).

INDEX

Achuar Indians, 142
Acre, Brazil, 60–1, 82–4
adaptation funds, 18, 48, 70, 197
Adeco Agropecuaria Brasil, 169
Africa, 8, 30–1, 35, 40, 42, 111, 139, 160, 187–8
agribusiness, 4, 17, 77, 91, 95–7, 99, 1045, 114–15, 118, 146, 155–7, 167, 170, 195–7
agricultural displacement, 164
aguaje tree, 182–3, 197–8
air pollution, 103
Alaska, 8
Alberto de Castro, Francisco, 81
alcool, 158
alpacas, 16, 23–5
aluvión, See glacial flood burst
Amazon Fund, 117–18
Amazon rainforest, 2–3, 40, 49–55, 60–3, 66, 69–70, 74–5, 87–8, 194–5
   and bats, 195
   and the "carbon bomb," 88
   chimneys, 74–5
   dieback theory, 54
   and drought, 55, 60–3, 66
   drying out of, 54, 75, 129
   essential role of, 49–53
   feedback loop, 75, 87
   and fruit, 197–8
   and medicinal drugs, 198
   as planetary air conditioner, 49–50
   reforestation, 136–7, 194–5
   size of, 3
   and temperature rise, 54–5, 69–70
   and wetlands, 87
   and world weather, 69

See carbon emissions; deforestation; forest fires; rainforest sinks; *terra preta*
Amazon River, 3, 53, 61, 75, 107
Amazon Watch, 143
Amazonia, 60
Amazonian Indians, 142, 179
Amazonian lowlands, 25
Amerindians, 59
amphibians, 64–6
Anapu, Brazil, 79–81
Ancash, Peru, 15, 19
anchoveta, 33, 35
anchovy, 33–4
Andean condor, 12
Andreas, Dwayne, 155
annatto tree, 75
Antarctica, 1, 110
AOL, 168
APL Jade, 133
apocalypse, 13
apus (sacred mountains), 12–13
arahuana fish, 181
archaeologists, 10, 12, 13, 15, 32–3, 51–3
Archer Daniels Midland (ADM), 97, 155, 167
Arctic Sunrise, 110–11
Argentina, 9, 26, 40, 76, 139, 165, 180, 188
Arhuaco Indians, 187, 200–1
Arias, Óscar, 191
Asia, 8, 30–1, 40
asthma, 103
Aguas Calientes, Peru, 15
Ausangate, Peru, 12–13